Lecture Notes in Computer Science   13150

More information about this series at https://link.springer.com/bookseries/558

Antonia Wachter-Zeh · Hannes Bartz ·
Gianluigi Liva (Eds.)

Code-Based Cryptography

9th International Workshop, CBCrypto 2021
Munich, Germany, June 21–22, 2021
Revised Selected Papers

Editors
Antonia Wachter-Zeh ⓘ
Technical University of Munich (TUM)
München, Germany

Gianluigi Liva ⓘ
Deutsches Zentrum für Luft-
und Raumfahrt (DLR)
Weßling, Germany

Hannes Bartz ⓘ
Deutsches Zentrum für Luft-
und Raumfahrt (DLR)
Weßling, Germany

ISSN 0302-9743   ISSN 1611-3349   (electronic)
Lecture Notes in Computer Science
ISBN 978-3-030-98364-2   ISBN 978-3-030-98365-9   (eBook)
https://doi.org/10.1007/978-3-030-98365-9

This Springer imprint is published by the registered company Springer Nature Switzerland AG
The registered company address is: Gewerbestrasse 11, 6330 Cham, Switzerland

Preface

Post-quantum code-based cryptography is a rapidly-developing research area lying at the intersection of several disciplines, including (algebraic) coding theory, information theory, and theoretical computer science. The topic gained momentum thanks to the recent effort put in place by the US National Institute of Standards and Technology (NIST) to standardize future-proof quantum-resistant cryptosystems. Originally named the "Code-Based Cryptography (CBC) Workshop", the series was initiated in 2009 as an informal forum with the goal of bringing together researchers active in the analysis and development of code-based encryption and authentication schemes. Over the years, the workshop has grown from a Europe-based, regional event to become a worldwide venue for the code-based cryptography community. The workshop was renamed "CBCrypto" in 2020, and its organization was extended to include the publication of selected manuscripts in the form of a post-proceedings volume. The 2021 edition of CBCrypto was originally planned to be held in Munich. Due to restrictions caused by the ongoing COVID-19 pandemic, the event was changed to a purely online format. Nevertheless, it proved to be a great success, with over 200 registered participants attending the talks from around the world.

The program was enriched by 14 contributed talks presenting recent research and works in progress. This book collects the six contributions that were selected for publication by the Program Committee through careful peer review. These contributions span all aspects of code-based cryptography, from design to implementation, including studies of security, new systems, and improved decoding algorithms. As such, the works presented in this book provide a synthesized yet significant overview of the state of the art of code-based cryptography, laying out the groundwork for future developments. We wish to thank the Program Committee members and the external reviewers for their hard and timely work.

November 2021

Antonia Wachter-Zeh
Hannes Bartz
Gianluigi Liva

Organization

Organizing Committee

Antonia Wachter-Zeh	Technical University of Munich (TUM), Germany
Hannes Bartz	German Aerospace Center (DLR), Germany
Gianluigi Liva	German Aerospace Center (DLR), Germany

Program Committee

Marco Baldi	Università Politecnica delle Marche, Italy
Gustavo Banegas	Technische Universiteit Eindhoven, The Netherlands
Alessandro Barenghi	Politecnico di Milano, Italy
Emanuele Bellini	Technology Innovation Institute, UAE
Sergey Bezzateev	Saint Petersburg University of Aerospace Instrumentation, Russia
Olivier Blazy	Université de Limoges, France
Pierre-Louis Cayrel	Laboratoire Hubert Curien, France
Franco Chiaraluce	Università Politecnica delle Marche, Italy
Alain Couvreur	LIX, École Polytechnique, France
Jean-Christophe Deneuville	École Nationale de l'Aviation Civile, France
Taraneh Eghlidos	Sharif University of Technology, Iran
Philippe Gaborit	University of Limoges, France
Cheikh T. Gueye	University of Dakar, Senegal
Anna-Lena Horlemann-Trautmann	University of St. Gallen, Switzerland
Pierre Loidreau	Université de Rennes, France
Chiara Marcolla	Technology Innovation Institute, UAE
Giacomo Micheli	University of South Florida, USA
Kirill Morozov	University of North Texas, USA
Ayoub Otmani	University of Rouen Normandie, France
Gerardo Pelosi	Politecnico di Milano, Italy
Edoardo Persichetti	Florida Atlantic University, USA
Joachim Rosenthal	University of Zurich, Switzerland
Simona Samardjiska	Radboud University, The Netherlands
Paolo Santini	Università Politecnica delle Marche, Italy
Nicolas Sendrier	Inria, France
John Sheekey	University College Dublin, Ireland
Jean-Pierre Tillich	Inria, France
Øyvind Ytrehus	University of Bergen, Norway

Additional Reviewers

Jessica Bariffi

Felicitas Hörmann

Thomas Jerkovits

Karan Khathuria

Francisco Lazaro

Julia Lieb

Balazs Matuz

Davide Orsucci

Cornelia Ott

Sabine Pircher

Sven Puchinger

Tania Richmond

Vladimir Sidorenko

Sebastian Stern

Violetta Weger

Contents

A Rank Metric Code-Based Group Signature Scheme

Olivier Blazy[✉], Philippe Gaborit[✉], and Dang Truong Mac[✉]

University of Limoges, Limoges, France
{olivier.blazy,gaborit}@unilim.fr, dang.mac@etu.unilim.fr

Abstract. Group signature is a major tool in today's cryptography. Rank based cryptography has been known for almost 30 years and recently reached the second round of the NIST competition for post-quantum primitives. In this work, we present a code-based group signature scheme in the rank metric context. The scheme follows the path presented by Ezerman *et al.* (ASIACRYPT' 2015) for Hamming metric but in a rank metric context which requires some specific adaptation and generalization. The scheme used a rank metric variation of the Stern's authentication scheme and relies solely on generic decoding problems. It also satisfies the CPA-anonymity and traceability properties in the random oracle model. In general the parameters of our scheme are slightly better compared to the Hamming scheme.

Keywords: Group signature · Code-based · Rank metric

1 Introduction

Designing group signature is one of the most intriguing problem in cryptography. The ultimate goal is schemes which satisfy the fundamental requirements of a group signature scheme, and meet practicable purposes. Especially that the era of quantum computing is coming, which would make number-theoretic based group signature schemes insecure, the search for post-quantum schemes has become active than ever. Much of proposals are published in both lattice-based assumptions and code-based assumptions.

On the lattice-based side, there have schemes such as [14,15]. Theoretically, these schemes provide efficient public key size and signature size in the asymptotic sense. (The size of signature only is linear in logarithm of the number of users.) However, as pointed out in [9], when being instantiated with practical parameters, they suffer from large key and signature sizes.

On the code-based side, there have constructions on both static and dynamic group, *e.g.,* [3,4,9]. The scheme presented in [4] used the RankSign primitive. However, RankSign signature scheme was broken in [11]. Thus, actually, there is no group signature scheme based on rank metric. The first static code-based group signature scheme in Hamming metric was designed by Ezerman *et al.* [9]. This scheme provides public key and signature sizes being linear in the number

© Springer Nature Switzerland AG 2022
A. Wachter-Zeh et al. (Eds.): CBCrypto 2021, LNCS 13150, pp. 1–21, 2022.
https://doi.org/10.1007/978-3-030-98365-9_1

2   O. Blazy et al.

of users which really is a weak point compared to those of lattice-based. Despite this fact, at the same level of security ($\lambda = 80$), their parameters are remarkably smaller than those of lattice schemes. Take a closer look, their construction uses 3 cryptographic layers:

1. The first layer is a signature scheme derived from Stern's identification protocol through Fiat-Shamir transform.
2. The second layer is the randomized McEliece encryption scheme which is used to encrypt identity of the signer.
3. The third layer is a zero-knowledge (ZK) protocol that links the two above layers together. It allows one to show that a given signature is generated by a certain user in the group who honestly encrypts his identity information.

It was emphasized also in their paper that "Constructing such protocol is quite challenging".

Our Contribution. In this work, we revisit that challenge and show how to adapt it in the rank context. Since [4] is broken, our scheme becomes the first rank metric group signature scheme, which, moreover, relies on generic problems. When being instantiated with concrete parameters at the same security level $\lambda = 128$, the size of signatures of our scheme are smaller than those of [9]; and when the values of ℓ are not too large, e.g., $\ell = 4, 8, 12$, the size of public keys are also less than those of [9]. Our parameters are set as in Sect. 5. For the schemes of [9] to attain security level 128, we take $(n, k, t) = (2^{12}, 3604, 41)$ as in [12], and for the Syndrome Decoding problem, we try to set $(m, r, w) = (4097, 721, 162)$ so that it also satisfies Lemma 1 in [9] (Table 1).

Table 1. Comparison with [9].

ℓ	PK size		Signature size	
	Our scheme	[9]	Our scheme	[9]
4	16.71 KB	2.22 MB	2.94 MB	3.05 MB
8	77.68 KB	2.24 MB	2.95 MB	3.05 MB
12	1.05 MB	2.58 MB	3.06 MB	3.12 MB
16	16.57 MB	8.12 MB	4.74 MB	4.85 MB

Overview of Our Techniques. Let $k, \ell, m, m_0, n, n_0, r_0, w_r, w_s$ be positive integers. We consider a group of $N - 1 = q^\ell - 1$ users. (The reason for this way of denoting will be clear in the sequel.) Each user is indexed by an integer $j \in \{1, \ldots, N-1\}$ and has a signing key $\mathbf{s}_j$, which is randomly chosen from $\mathcal{S}_{w_s}^{n_0, m_0}$, i.e., the set of vectors of rank wight w_s in $\mathbb{F}_{q^{m_0}}^{n_0}$. A part of public key contains a matrix $\mathbf{H} \in \mathbb{F}_{q^{m_0}}^{r_0 \times n_0}$, which is the parity matrix in the systematic form of an ideal code, and $N-1$ syndromes $\mathbf{y}_1, \ldots, \mathbf{y}_{N-1} \in \mathbb{F}_{q^m}^{r_0}$ such that $\mathbf{H} \cdot \mathbf{s}_j^T = \mathbf{y}_j^T$. Our three layers are as follows.

1. **The signature layer.** Let $\mathbf{A} = [\mathbf{y}_1^T | \cdots | \mathbf{y}_{N-1}^T] \in \mathbb{F}_{q^{m_0}}^{r_0 \times (N-1)}$, and $\mathbf{x} = \delta_j^{N-1}$ - the vector of dimension $N-1$ with 1 at the jth position and 0 elsewhere. In this layer, the user uses Stern's framework in the rank context to prove that he possesses a pair $(\mathbf{s}, \mathbf{x})$ satisfying

$$\mathbf{H} \cdot \mathbf{s}^T - \mathbf{A} \cdot \mathbf{x}^T = \mathbf{0}. \tag{1}$$

Then, the protocol is transformed into a Fiat-Shamir signature.

2. **The encryption layer.** We use RQC encryption scheme to encrypt the identity information of the users. Each index $j \in \{1, \ldots, N-1\}$ is mapped to a vector in $\mathbb{F}_{q^m}^k$, the message space of an RQC scheme, by function $\mathsf{I2V}(\cdot)$. The ciphertext is of the form $\mathbf{c} = (\mathbf{c}_1, \mathbf{c}_2)$ such that

$$\begin{cases} \mathbf{c}_1 = \mathbf{r}_1 + \mathbf{h} \cdot \mathbf{r}_2, \\ \mathbf{c}_2 = \mathbf{r}_3 + \mathbf{s} \cdot \mathbf{r}_2 + \mathsf{I2V}(j) \cdot \mathbf{G}, \end{cases} \tag{2}$$

where $\mathbf{h}, \mathbf{s} \in \mathbb{F}_{q^m}^n$ is the public key, $\mathbf{G} \in \mathbb{F}_{q^m}^k$ is a generator matrix of a public code, and $\mathbf{r}_i \in \mathcal{S}_{w_r}^{n,m}$ for $i = 1, 2, 3$.

3. **The third layer.** This ingredient is a ZK protocol that allows the user to show that the vector $\mathbf{x} = \delta_j^{N-1}$ used in the first layer and the hidden plaintext $\mathsf{I2V}(j)$ used in the second layer both point to the same $j \subset [1, \ldots, N-1]$. Our solution to this problem is as follows. Let $f_1 \colon \{1, \ldots, N-1\} \to \mathbb{F}_q^\ell$ be the function that maps each element of $\{1, \ldots, N-1\}$ to a different element of $\mathbb{F}_q^\ell$, $f_2 \colon \mathbb{F}_q^\ell \to \mathbb{F}_{q^m}^k$ be a function that maps each vector in $\mathbb{F}_q^\ell$ to a vector in $\mathbb{F}_{q^m}^k$; their inverses are denoted by f_1^{-1}, f_2^{-1}, respectively. The map $\mathsf{I2V}$ is defined as $\mathsf{I2V}(j) = f_2 \circ f_1(j)$. For every $\mathbf{v} \in \mathbb{F}_q^\ell \setminus \{\mathbf{0}\}$, we construct two permutations $P_{\mathbf{v}} \colon \mathbb{F}_q^{N-1} \to \mathbb{F}_q^{N-1}$ and $P'_{\mathbf{v}} \colon S \to S$, where $S = \{\mathsf{I2V}(j) \mid j \in \{1, \ldots, N-1\}\} \cup \{\mathbf{0}\}$, such that for any $j \in \{1, \ldots, N-1\}$ we have

$$\mathbf{x} = \delta_j^{N-1} \Longleftrightarrow P_{\mathbf{v}}(\mathbf{x}) = \delta_{f_1^{-1}(f_1(j) \cdot \mathbf{v})}^{N-1},$$

$$\mathbf{m} = \mathsf{I2V}(j) \Longleftrightarrow P'_{\mathbf{v}}(\mathbf{m}) = \mathsf{I2V}\left(f_1^{-1}(f_1(j) \cdot \mathbf{v})\right).$$

In the protocol, the user randomly picks a non zero vector $\mathbf{v} \in \mathbb{F}_q^\ell \setminus \{\mathbf{0}\}$ and sends $\mathbf{v}' = f_1(j) \cdot \mathbf{v}$. The verifier, seeing that $P_{\mathbf{v}}(\mathbf{x}) = \delta_{f_1^{-1}(\mathbf{v}')}^{N-1}$ and $P'_{\mathbf{v}}(\mathbf{m}) = \mathsf{I2V}(f_1^{-1}(\mathbf{v}'))$, should be convinced that $\mathbf{x}$ and $\mathbf{m}$ link to the same $j \in \{1, \ldots, N-1\}$, yet the value of j is completely hidden. Here, vector $\mathbf{v}$ also acts as a one-time pad as in the case of addition, *i.e.*, the operation $\oplus/+$. We remark that our method can be applied very well in the case of [9]. Note that the technique used in [9] heavily relies on the particular value $q = 2$. When the binary field is replaced by any finite field, this technique would lose its efficiency.

With this technique embedded in Stern's framework, the user can convince the verifier that he possesses a tuple $(j, \mathbf{s}, \mathbf{x}, \mathbf{r}_1, \mathbf{r}_2, \mathbf{r}_3)$ satisfying 1 and 2. By repeating this protocol many times, and making non-interactive, we get a ZK proof of knowledge Π. The final signature is of the form $(\mathbf{c}, \Pi)$. In the random oracle model, the anonymity of the scheme relies on the zero-knowledge of Π and the CPA-security of the RQC; its traceability relies on the hardness of Rank Syndrome Decoding problem.

2 Preliminaries

2.1 Notations

We use bold low-case letters to denote vectors and bold capital letters for matrices. The transpose of a vector $\mathbf{x}$ is denoted by $\mathbf{x}^T$. The same notation is used for matrices. The rank weight of a vector $\mathbf{x}$ is denoted by $\|\mathbf{x}\|$. The set of invertible matrices of size m over $\mathbb{F}_q$ is denoted by $\mathrm{GL}(m, q)$. For a positive integer $N > 1$, $[N-1] \overset{\text{def}}{=} \{1, \ldots, N-1\}$. The sphere of radius r centered at $\mathbf{0}$ in $\mathbb{F}_{q^m}^n$ is denoted by $\mathcal{S}_r^{n,m}$. By writing $x \leftarrow X$, we mean that x is drawn according to the distribution X, if X is a distribution; or drawn uniformly at random from X when X is a set; or output of the algorithm X, if X is an algorithm.

2.2 Background on Code-Based Cryptography

Let m, n be two positive integers and q be a power of a prime number. Let $\{\alpha_1, \ldots, \alpha_m\}$ be a basis of $\mathbb{F}_{q^m}$ over $\mathbb{F}_q$. This basis can be used to associate any vector $\mathbf{x} = (x_1, \ldots, x_n) \in \mathbb{F}_{q^m}^n$ to the corresponding matrix $\mathbf{A_x} \in \mathbb{F}_q^{n \times m}$ as

$$\begin{pmatrix} x_1 \\ \vdots \\ x_n \end{pmatrix} = \begin{pmatrix} a_{11} & \cdots & a_{1m} \\ \vdots & \vdots & \vdots \\ a_{n1} & \cdots & a_{nm} \end{pmatrix} \cdot \begin{pmatrix} \alpha_1 \\ \vdots \\ \alpha_m \end{pmatrix}.$$

The rank weight of $\mathbf{x}$ is defined to be the rank of matrix $\mathbf{A_x}$, that is, $\|\mathbf{x}\| := \mathrm{rank}(\mathbf{A_x})$. In this metric, the distance between two vectors $\mathbf{x}$ and $\mathbf{y}$, denoted by $d(\mathbf{x}, \mathbf{y})$, is defined to be equal the rank weight of $\mathbf{x} - \mathbf{y}$, i.e., $d(\mathbf{x}, \mathbf{y}) := \|\mathbf{x} - \mathbf{y}\|$.

Now, let $f(X) \in \mathbb{F}_{q^m}[X]$ be a polynomial of degree n and $\mathbb{F}_{q^m}[X]/\langle f \rangle = \mathcal{R}_f$. Consider the following map:

$$\phi \colon \mathbb{F}_{q^m}^n \longrightarrow \mathcal{R}_f$$
$$(a_0, \ldots, a_{n-1}) \longmapsto a_0 + \cdots + a_{n-1}X^{n-1}.$$

The inverse map, denoted by ϕ^{-1}, simply maps a polynomial to the vector formed by its coefficients. For the sake of simplicity, if $\mathbf{a} = (a_0, \ldots, a_{n-1}) \in \mathbb{F}_{q^m}^n$, we let $\phi(\mathbf{a}) = a_0 + \cdots + a_{n-1}X^{n-1} = a(X)$. For $\mathbf{a}, \mathbf{b} \in \mathbb{F}_{q^m}$, their product $\mathbf{a} \cdot \mathbf{b}$ is defined as

$$\mathbf{a} \cdot \mathbf{b} = \phi^{-1}\big(a(X) \cdot b(X)\big).$$

Clearly, we have $\mathbf{a} \cdot \mathbf{b} = \mathbf{b} \cdot \mathbf{a}$. It is also not hard to see that

$$\mathbf{a} \cdot \mathbf{b} = (a_0, \ldots, a_{n-1}) \cdot \begin{pmatrix} \phi^{-1}\big(b(X)\big) \\ \vdots \\ \phi^{-1}\big(X^{n-1}b(X)\big) \end{pmatrix}. \tag{3}$$

Consider the case when $f(X)$ is irreducible over $\mathbb{F}_{q^m}$, then $g(X)$ and $f(X)$ are coprime for any nonzero $g(X) \in \mathcal{R}_f$. Thus, for an arbitrary nonzero $g(X) \in \mathcal{R}_f$, if we define

$$g \cdot \mathcal{R}_f = \{g(X) \cdot a(X) \mod f \mid a(X) \in \mathcal{R}_f\},$$

then we have $g \cdot \mathcal{R}_f = \mathcal{R}_f$. From this observation, we deduce that $\mathbf{g} \cdot \mathbb{F}_{q^m}^n = \mathbb{F}_{q^m}^n$ (here, $\mathbf{g} = \phi^{-1}(g) \neq \mathbf{0}$), that is to say, $\mathbf{g}$ together with the multiplication operation defined above form a permutation over $\mathbb{F}_{q^m}^n$. (This permutation fixes $\mathbf{0}$.) For our purpose, it is enough to consider $m = 1$ and $f(X)$ is irreducible over $\mathbb{F}_q$.

The right-most term on the right hand side of Eq. 3 is usually referred to as the ideal matrix generated by $b(X)$ with respect to $f(X)$. For ease of notation, vectors are identical with their corresponding polynomials, i.e., $X^k \mathbf{b}$ is understood to be $\phi^{-1}(X^k b(X))$. Thus, the ideal matrix of a vector $\mathbf{b}$ with respect to f is written as

$$\mathfrak{b} = \begin{pmatrix} \mathbf{b} \\ X \cdot \mathbf{b} \\ \vdots \\ X^{n-1} \cdot \mathbf{b} \end{pmatrix}.$$

In our construction, we will use 2-ideal codes and 3-ideal codes. A 2-ideal code of length $2n$ over $\mathbb{F}_{q^m}$ is a code whose parity matrix is of the form

$$\mathbf{H} = [\mathbf{I}_n \mid \mathfrak{h}^T], \tag{4}$$

where $\mathfrak{h}$ is the ideal matrix of a vector $\mathbf{h}$ in $\mathbb{F}_{q^m}^n$. Similarly, a 3-ideal code of length $3n$ over $\mathbb{F}_{q^m}$ is a code whose parity matrix is of the form

$$\mathbf{H} = \begin{pmatrix} \mathbf{I}_n & \mathbf{0} & \mathfrak{h}_1^T \\ \mathbf{0} & \mathbf{I}_n & \mathfrak{h}_2^T \end{pmatrix}. \tag{5}$$

For a given vector $\mathbf{x} \in \mathbb{F}_{q^m}^n$, we associate with it the vector space defined by its coordinates.

Definition 1. *Let* $\mathbf{x} = (x_1, \ldots, x_n) \in \mathbb{F}_{q^m}^n$. *The vector space over* $\mathbb{F}_q$ *defined by* $x_1, \ldots, x_n$ *is called the support of* $\mathbf{x}$, *and denoted by* $\mathrm{Supp}(\mathbf{x})$. *That is,*

$$\mathrm{Supp}(\mathbf{x}) = \mathrm{Span}_{\mathbb{F}_q}(x_1, \ldots, x_n).$$

Next, we recall some definitions concerning code-based hardness assumptions.

Definition 2 (Rank Syndrome Decoding Problem). *Let* $n, k,$ *and* w *be positive integers,* $\mathbf{H}$ *a random matrix over* $\mathbb{F}_{q^m}^{(n-k) \times n}$, *and* $\mathbf{y}$ *a random vector in* $\mathbb{F}_{q^m}^{n-k}$. *The rank syndrome decoding problem,* $\mathrm{RSD}(n, k, w)$, *asks to find a vector* $\mathbf{x} \in \mathcal{S}_w^{n,m}$ *such that* $\mathbf{H}\mathbf{x}^T = \mathbf{y}^T$.

Definition 3 (Rank Syndrome Decoding Distribution). *Let* $n, k,$ *and* w *be positive integers, the* $\mathrm{RSD}(n, k, w)$ *distribution chooses* $\mathbf{H} \leftarrow \mathbb{F}_{q^m}^{(n-k) \times n}$ *and* $\mathbf{x} \leftarrow \mathcal{S}_w^{n,m}$, *and outputs* $(\mathbf{H}, \mathbf{H} \cdot \mathbf{x}^T)$.

Definition 4 (Decisional Rank Syndrome Decoding Problem). *The decisional RSD problem, DRSD(n, k, w), asks to decide with non-negligible advantage whether an instance $(\mathbf{H}, \mathbf{y}^T)$ came from the RSD(n, k, w) distribution or the uniform distribution over $\mathbb{F}_{q^m}^{(n-k) \times n} \times \mathbb{F}_{q^m}^{n-k}$.*

In the following definitions, $\nu \in \{2, 3\}$ and $S(n, \nu)$ is the set of all matrices of the form as in Eq. 4 or 5 corresponding to the case $\nu = 2$ or $\nu = 3$, respectively.

Definition 5 (ν − IRSD Distribution). *Let n, w be positive integers, $P(X) \in \mathbb{F}_q[X]$ be an irreducible polynomial of degree n. The ν − IRSD(n, w) distribution chooses uniformly at random a matrix $\mathbf{H} \in S(n, \nu)$ together with a vector $\mathbf{x} \in \mathbb{F}_{q^m}^{\nu n}$ such that $\|\mathbf{x}\| = w$ and outputs $(\mathbf{H}, \mathbf{H} \cdot \mathbf{x}^T)$.*

Definition 6 (Computational ν − IRSD Problem). *Let n, w be positive integers, $P(X) \in \mathbb{F}_q[X]$ be an irreducible polynomial of degree n, $\mathbf{H} \in S(n, \nu)$ be a random matrix, and $\mathbf{y} \leftarrow \mathbb{F}_{q^m}^n$. The computational ν − IRSD(n, w) problem asks to find a vector $\mathbf{x} \in \mathbb{F}_{q^m}^{\nu n}$ such that $\|\mathbf{x}\| = w$ and $\mathbf{H} \cdot \mathbf{x}^T = \mathbf{y}^T$.*

Definition 7 (Decisional ν − IRSD Problem). *The decisional ν − IRSD(n, w) problem asks to decide with non-negligible advantage whether $(\mathbf{H}, \mathbf{y}^T)$ came from the ν − IRSD(n, w) distribution or the uniform distribution over $S(n, \nu) \times \mathbb{F}_{q^m}^n$.*

The RQC Scheme. In the Encryption layer, we make use of the RQC scheme [2]. It is as follows.

- RQC.Setup(1^λ): Generate parameters $m = m(\lambda), n = n(\lambda), k = k(\lambda), w_r = w_r(\lambda)$, an irreducible polynomial $P[X] \in \mathbb{F}_q[X]$, which is also irreducible in $\mathbb{F}_{q^m}[X]$. The plaintext space is $\mathbb{F}_{q^m}^k$. Output param $= (m, n, k, w_r, P)$.
- RQC.KeyGen(param): Generate $\mathbf{h} \leftarrow \mathbb{F}_{q^m}^n, \mathbf{x}, \mathbf{y} \leftarrow \mathcal{S}_{w_r}^{n,m}$ sharing the same support, a generator matrix $\mathbf{G} \in \mathbb{F}_{q^m}^{k \times n}$ of a public code $\mathcal{C}$. Output $\mathsf{pk_{RQC}} = (\mathbf{h}, \mathbf{s} = \mathbf{x} + \mathbf{h} \cdot \mathbf{y}, \mathbf{G})$ and $\mathsf{sk_{RQC}} = (\mathbf{x}, \mathbf{y})$.
- RQC.Enc($\mathsf{pk_{RQC}}, \mathbf{m}$): To encrypt a message $\mathbf{m} \in \mathbb{F}_{q^m}^k$, choose $\mathbf{r}_1, \mathbf{r}_2, \mathbf{r}_3 \leftarrow \mathcal{S}_{w_r}^{n,m}$, which belong to the same support. Compute

$$\begin{cases} \mathbf{c}_1 = \mathbf{r}_1 + \mathbf{h} \cdot \mathbf{r}_2, \\ \mathbf{c}_2 = \mathbf{s} \cdot \mathbf{r}_2 + \mathbf{r}_3 + \mathbf{m} \cdot \mathbf{G}. \end{cases}$$

- RQC.Dec($\mathsf{sk_{RQC}}, \mathbf{c}$): Apply the decoding algorithm of the code $\mathcal{C}$ to

$$\mathbf{y} \cdot \mathbf{c}_1 - \mathbf{c}_2 = \mathbf{x} \cdot \mathbf{r}_2 - \mathbf{y} \cdot \mathbf{r}_1 + \mathbf{r}_3 + \mathbf{m} \cdot \mathbf{G}.$$

For the sake of convenience, define

$$\mathbf{H}_1 = \begin{pmatrix} \mathbf{I}_n \\ \mathbf{0} \end{pmatrix}, \quad \mathbf{H}_2 = \begin{pmatrix} \mathfrak{h}^T \\ \mathfrak{s}^T \end{pmatrix}, \quad \mathbf{H}_3 = \begin{pmatrix} \mathbf{0} \\ \mathbf{I}_n \end{pmatrix}, \quad \mathbf{H}_4 = \begin{pmatrix} \mathbf{0} \\ \mathbf{G}^T \end{pmatrix},$$

then we have $[\mathbf{H}_1 | \cdots | \mathbf{H}_4] \cdot (\mathbf{r}_1, \mathbf{r}_2, \mathbf{r}_3, \mathbf{m})^T = (\mathbf{c}_1, \mathbf{c}_2)^T$. The RQC scheme is CPA-secure; its security relies on the hardness of the decisional 2 − IRSD(n, w_r) and 3 − IRSD(n, w_r) problems as has been proven in [2]. (Although the proof therein is applied for quasi-cyclic codes, a proof for ideal codes can be derived straightforwardly).

2.3   Group Signatures

In this section, we recall some definitions of group signatures following [7] on the case of static groups.

Definition 8. *A group signature scheme* $\mathcal{GS}$ = (KeyGen, Sign, Verify, Open) *contains four polynomial-time algorithms:*

1. KeyGen *takes as input* $(1^\lambda, 1^N)$, *where* λ *is the security parameter and* N *is a positive integer which is the number of group users, and returns a tuple* (gpk, gmsk, gsk), *where* gpk *is the group public key,* gmsk *is the group manager's secret key, and* gsk = $\{$gsk$[j]\}_{j\in[N-1]}$ *with* gsk$[j]$ *being the secret key of the group user of index* j.
2. Sign *takes as input a message* M, *a secret key* gsk$[j]$ *in the set* gsk *and returns a group signature* Σ *on* M.
3. Verify *takes as input the group public key* gpk, *a message* M, *a signature* Σ *on* M, *and returns either* 1 (Accept) *or* 0 (Reject).
4. Open *takes as input the group manager's secret key* gmsk, *a signature* M, *a signature* Σ *on* M, *and returns an identity* j *or the symbol* $\perp$ *to indicate failure.*

Correctness: The correctness of a group signature scheme requires that for all positive integers λ, N, all output (gpk, gmsk, gsk) of KeyGen, all identity j, and all message $M \in \{0,1\}^*$,

$$\begin{cases} \text{Verify}(\text{gpk}, M, \text{Sign}(\text{gsk}[j], M)) = 1, \\ \text{Open}(\text{gmsk}, M, \text{Sign}(\text{gsk}[j], M)) = j. \end{cases}$$

Security Notions: A secure group signature scheme must satisfy two security requirements:

1. *Traceability* requires that all signatures can be traced back to the identity of its signer, even in the case there is a collusion between the group users.
2. *Anonymity* requires that signatures generated by two users are computationally indistinguishable to an adversary who knows all the secret keys.

We follow [9] by stating the security definitions.

Definition 9. *A group signature scheme* $\mathcal{GS}$ = (KeyGen, Sign, Verify, Open) *is CPA-anonymous if for all polynomial* $N(\cdot)$ *and any* PPT *adversary* $\mathcal{A}$, *the advantage of* $\mathcal{A}$ *in the following experiment is negligible in* λ:

1. *Run* (gpk, gmsk, gsk) $\leftarrow$ KeyGen$(1^\lambda, 1^N)$ *and send* (gpk, gsk) *to* $\mathcal{A}$.
2. $\mathcal{A}$ *outputs two identities* $j_0, j_1 \in [N-1]$ *together with a message* M. *Choose a random bit* b *and give* Sign(gsk$[j_b], M)$ *to* $\mathcal{A}$. *Then* $\mathcal{A}$ *outputs a bit* b'.

$\mathcal{A}$ *succeeds if* $b' = b$. *The advantage of* $\mathcal{A}$ *is defined to equal* $\left| \Pr[\mathcal{A} \text{ succeeds}] - \frac{1}{2} \right|$.

Definition 10. *A group signature* $\mathcal{GS} = (\mathsf{KeyGen}, \mathsf{Sign}, \mathsf{Verify}, \mathsf{Open})$ *is traceable if for all polynomial* $N(\cdot)$ *and any adversary* $\mathcal{A}$, *the success probability of* $\mathcal{A}$ *in the following experiment is negligible in* λ:

1. *Run* $(\mathsf{gpk}, \mathsf{gmsk}, \mathsf{gsk}) \leftarrow \mathsf{KeyGen}(1^\lambda, 1^N)$ *and send* $(\mathsf{gpk}, \mathsf{gsk})$ *to* $\mathcal{A}$.
2. $\mathcal{A}$ *may query the following oracles adaptively and in any order:*
   - *An* $\mathcal{O}^{\mathsf{Corrupt}}$ *oracle that on input* $j \in [N-1]$, *outputs* $\mathsf{gsk}[j]$.
   - *An* $\mathcal{O}^{\mathsf{Sign}}$ *oracle that on input* j *and a message* M, *returns* $\mathsf{Sign}(\mathsf{gsk}[j], M)$.
   Let CU *be the set of identities queried to* $\mathcal{O}^{\mathsf{Corrupt}}$.
3. $\mathcal{A}$ *outputs a message* M^* *and a signature* Σ^*.

$\mathcal{A}$ *succeeds if* $(i)\, \mathsf{Verify}(\mathsf{gpk}, M^*, \Sigma^*) = 1$ *and* $(ii)\, \mathsf{Sign}(\mathsf{gsk}[j], M^*)$ *was never queried for* $j \notin CU$, *and yet* $(iii)\, \mathsf{Open}(\mathsf{gmsk}, M^*, \Sigma^*) \notin CU$.

2.4   Transform of Index

Let $N - 1 = q^\ell - 1$ be the number of users, $\mathcal{B} = \{\alpha_1, \ldots, \alpha_m\}$ be a basis for $\mathbb{F}_{q^m}$ over $\mathbb{F}_q$, $p(X)$ be an irreducible polynomial of degree ℓ over $\mathbb{F}_q$, and $\mathbf{B} \in \mathbb{F}_q^{\ell \times mk}$ be a generator matrix of the systematic form of some q-ary linear code $\mathcal{C}$. We define a map $\mathsf{I2V} \colon [N-1] \longrightarrow \mathbb{F}_{q^m}^k$ as follows.

1. $f_1 \colon [N-1] \longrightarrow \mathbb{F}_q^\ell$ is any public injective map such that $f_1(j) \neq \mathbf{0}$. For example, let α a primitive element of $\mathbb{F}_q$, i.e., $\mathbb{F}_q = \{0, \alpha, \ldots, \alpha^{q-1}\}$, and $f \colon [q-1] \cup \{0\} \to \mathbb{F}_q$ the map such that $f(0) = 0$ and $f(i) = \alpha^i$ for $i = 1, \ldots, q-1$, then $f_1(j) = \big(f(x_0), \ldots, f(x_{\ell-1})\big)$, where $j = x_0 + \cdots + x_{\ell-1} q^{\ell-1}$ is the representation of j in the base q.
2. $f_2 \colon \mathbb{F}_q^\ell \longrightarrow \mathbb{F}_{q^m}^k$ defined as follows: for a vector $(a_0, \ldots, a_{\ell-1}) \in \mathbb{F}_q^\ell$, compute

$$(b_0, \ldots, b_{mk-1}) = (a_0, \ldots, a_{\ell-1}) \cdot \mathbf{B}$$

   and form the matrix

$$\mathbf{A} = \begin{pmatrix} b_0 & \cdots & b_{m-1} \\ \vdots & \vdots & \vdots \\ b_{(k-1)m} & \cdots & b_{mk-1} \end{pmatrix}.$$

   Then

$$f_2(a_0, \ldots, a_{\ell-1}) := (\alpha_1, \ldots, \alpha_m) \cdot \mathbf{A}^T.$$

3. Define $\mathsf{I2V}(j) := f_2 \circ f_1(j)$, where $\circ$ denotes the composition of mapping.

Let S denote the image of $\mathsf{I2V}$, then S is a subset of cardinality N of $\mathbb{F}_{q^m}^k$. Conversely, for each vector $\mathbf{m} = (m_1, \ldots, m_k) \in S$, there is a unique $j \in [N-1] \cup \{0\}$ such that $\mathsf{I2V}(j) = \mathbf{m}$. (If $\mathbf{m} = \mathbf{0}$, then j is set to be equal to 0.) The inverse map is denoted by $\mathsf{V2I} := f_1^{-1} \circ f_2^{-1}$.

2.5 Permutations

Let $\mathbf{v} \in \mathbb{F}_q^\ell \backslash \{\mathbf{0}\}$ be a random vector. We define two permutations:

- $P_{\mathbf{v}} : \mathbb{F}_q^{N-1} \longrightarrow \mathbb{F}_q^{N-1}$ transforms $\mathbf{x} = (x_1, \ldots, x_{N-1})$ to $\mathbf{x}' = (x'_1, \ldots, x'_{N-1})$, where $x_i = x'_{f_1^{-1}(f_1(i) \cdot \mathbf{v})}$. Here, the multiplication is defined with respect to $p(X)$. Therefore,

$$\mathbf{x} = \delta_j \Longleftrightarrow P_{\mathbf{v}}(\mathbf{x}) = \delta^{N-1}_{f_1^{-1}(f_1(j) \cdot \mathbf{v})}.$$

- $P'_{\mathbf{v}} : S \longrightarrow S$ as follows. For a vector $\mathbf{z} \in S$, let $\mathbf{z}_1 = f_2^{-1}(\mathbf{z}) \in \mathbb{F}_q^\ell$. Let $\mathbf{z}_2 = \mathbf{v} \cdot \mathbf{z}_1$ and define $P'_{\mathbf{v}}(\mathbf{z}) = f_2(\mathbf{z}_2)$. Since $f_2^{-1}(\mathsf{I2V}(j)) = f_1(j)$, so clearly,

$$\mathbf{m} = \mathsf{I2V}(j) \Longleftrightarrow P'_{\mathbf{v}}(\mathbf{m}) = f_2(f_1(j) \cdot \mathbf{v}).$$

An operation in the rank metric which is equivalent to the permutation notion in the Hamming metric was first introduced in [13]. We recall it here as follows. For a given basis $\mathcal{B}$ of $\mathbb{F}_{q^m}$ over $\mathbb{F}_q$, let $\varphi_\mathcal{B}$ be the map that associates the vectors in $\mathbb{F}_{q^m}^n$ to their corresponding matrices with respect to $\mathcal{B}$. Then, for an invertible matrix $\mathbf{Q}$ of size m over $\mathbb{F}_q$ and a vector $\mathbf{x} \in \mathbb{F}_{q^m}^n$, we define

$$\mathbf{Q} \star \mathbf{x} = \psi_\mathcal{B}^{-1}(\mathbf{A_x} \cdot \mathbf{Q}).$$

This operation allows us to transform a given vector to another of the same rank.

Lemma 1. *Let $\mathbf{Q} \in \mathrm{GL}(m, q)$ and $\mathbf{P}_1, \mathbf{P}_2 \in \mathrm{GL}(n, q)$. Then $\mathbf{x}, \mathbf{y} \in \mathbb{F}_{q^m}^n$ are in the same support if and only if $\mathbf{Q} \star \mathbf{x} \mathbf{P}_1$ and $\mathbf{Q} \star \mathbf{y} \mathbf{P}_2$ are in the same support.*

Proof. See [13].

3 The Underlying Interactive Protocol

3.1 The Interactive Scheme

This section is devoted to our zero-knowledge argument of knowledge. Let $k, \ell, m,$ $m_0, n, n_0, r_0, w_r, w_s$ be positive integers. The number of group users is $N - 1 = q^\ell - 1$. The common input contains matrices $\mathbf{H} \in \mathbb{F}_{q^{m_0}}^{r_0 \times n_0}, \mathbf{H}_1, \ldots, \mathbf{H}_4, N - 1$ syndromes $\mathbf{y}_1, \ldots, \mathbf{y}_{N-1} \in \mathbb{F}_{q^{m_0}}^{r_0}$, a ciphertext $\mathbf{c} = (\mathbf{c}_1, \mathbf{c}_2) \in \mathbb{F}_{q^m}^{2n}$, a basis $\mathcal{B} = \{\alpha_1, \ldots, \alpha_m\}$ of $\mathbb{F}_{q^m}$ over $\mathbb{F}_q$, a generator matrix $\mathbf{B} \in \mathbb{F}_q^{\ell \times mk}$ of a code $\mathcal{C}$, an irreducible polynomial $p(X)$ of degree ℓ over $\mathbb{F}_q[X]$, and the map f_1. The output of the protocol is that prover $\mathcal{P}$ simultaneously convinces verifier $\mathcal{V}$ in zero-knowledge that $\mathcal{P}$ possesses a vector $\mathbf{s} \in \mathcal{S}_{w_s}^{n_0, m_0}$ corresponding to a certain syndrome $\mathbf{y}_j \in \{\mathbf{y}_1, \ldots, \mathbf{y}_{N-1}\}$ with hidden index j, and that $\mathbf{c}$ is a correct encryption of $\mathbf{m} = \mathsf{I2V}(j)$ using the RQC scheme described by $\mathbf{H}_1, \ldots, \mathbf{H}_4$. More precisely, the secret witness of $\mathcal{P}$ is a tuple $(j, \mathbf{s}, \mathbf{r}_1, \mathbf{r}_2, \mathbf{r}_3) \in [N-1] \times \mathbb{F}_{q^{m_0}}^{n_0} \times \mathbb{F}_{q^m}^n \times \mathbb{F}_{q^m}^n \times \mathbb{F}_{q^m}^n$ such that

$$\begin{cases} \mathbf{H} \cdot \mathbf{s}^T = \mathbf{y}^T \quad \wedge \quad \mathbf{s} \in \mathcal{S}_{w_s}^{n_0, m_0}, \\ \widetilde{\mathbf{H}} \cdot \left(\mathbf{r}_1, \mathbf{r}_2, \mathbf{r}_3, \mathsf{I2V}(j)\right)^T = \mathbf{c}^T \quad \wedge \quad \mathbf{r}_i \in \mathcal{S}_{w_r}^{n, m}, i = 1, 2, 3, \end{cases}$$

where $\widetilde{\mathbf{H}} = [\mathbf{H}_1 | \cdots | \mathbf{H}_4]$. Let $\mathbf{A} = \left[\mathbf{y}_1^T | \cdots | \mathbf{y}_{N-1}^T\right] \in \mathbb{F}_{q^{m_0}}^{r_0 \times (N-1)}$, $\mathbf{m} = \mathsf{I2V}(j)$, and $\mathbf{x} = \delta_j^{N-1}$ be the index representation vector of j. Then, the above equations can be expressed as

$$\begin{cases} \mathbf{H} \cdot \mathbf{s}^T - \mathbf{A} \cdot \mathbf{x}^T = 0 \quad \wedge \quad \mathbf{x} = \delta_j^{N-1} \quad \wedge \quad \mathbf{s} \in \mathcal{S}_{w_s}^{n_0, m_0}, \\ \widetilde{\mathbf{H}} \cdot \left(\mathbf{r}_1, \mathbf{r}_2, \mathbf{r}_3, \mathbf{m}\right)^T = \mathbf{c}^T \quad \wedge \quad \mathbf{m} = \mathsf{I2V}(j) \quad \wedge \quad \mathbf{r}_i \in \mathcal{S}_{w_r}^{n, m}, i = 1, 2, 3. \end{cases}$$

A ZKAoK for the above relations is obtained as follows:

- To prove that $\mathbf{x} = \delta_j^{N-1}$ and $\mathbf{m} = \mathsf{I2V}(j)$ without revealing j, prover $\mathcal{P}$ randomly picks a vector $\mathbf{v} \in \mathbb{F}_q^\ell \setminus \{\mathbf{0}\}$, sends $\mathbf{v}' = f_1(j) \cdot \mathbf{v}$ and shows that

$$P_{\mathbf{v}}(\mathbf{x}) = \delta_{j'}^{N-1} \quad \text{and} \quad P'_{\mathbf{v}}(\mathbf{m}) = \mathsf{I2V}(j'),$$

  where $j' = f_1^{-1}(\mathbf{v}')$.
- To prove in zero-knowledge that $\mathbf{s} \in \mathcal{S}_{w_s}^{n_0, m_0}$, prover $\mathcal{P}$ chooses random matrices $\mathbf{Q}_0 \leftarrow \mathrm{GL}(m_0, q)$, $\mathbf{P}_0 \leftarrow \mathrm{GL}(n_0, q)$, and shows that $\mathbf{Q}_0 \star \mathbf{s} \mathbf{P}_0 \in \mathcal{S}_{w_s}^{n_0, m_0}$. To prove in zero-knowledge that $\mathbf{r}_i \in \mathcal{S}_{w_r}^{n, m}$ and that they share the same support, $\mathcal{P}$ samples randomly $\mathbf{Q} \leftarrow \mathrm{GL}(m, q), \mathbf{P}_i \leftarrow \mathrm{GL}(n, q)$ and shows that $\mathbf{Q} \star \mathbf{r}_i \mathbf{P}_i \in \mathcal{S}_{w_r}^{n, m}$ and have the same support, for $i = 1, 2, 3$. We refer the reader to [8] for more details.
- To prove the linear equations in ZK, $\mathcal{P}$ samples $(\mathbf{v_s}, \mathbf{v_x}, \mathbf{v}_1, \mathbf{v}_2, \mathbf{v}_3, \mathbf{v_m})$ randomly and shows that

$$\begin{cases} \mathbf{H} \cdot (\mathbf{s} + \mathbf{v_s})^T - \mathbf{A} \cdot (\mathbf{x} + \mathbf{v_x})^T = \mathbf{H} \cdot \mathbf{v_s}^T - \mathbf{A} \cdot \mathbf{v_x}^T, \\ \widetilde{\mathbf{H}} \cdot \left(\mathbf{r}_1 + \mathbf{v}_1, \mathbf{r}_2 + \mathbf{v}_2, \mathbf{r}_3 + \mathbf{v}_3, \mathbf{m} + \mathbf{v_m}\right)^T - \mathbf{c}^T = \widetilde{\mathbf{H}} \cdot \left(\mathbf{v}_1, \mathbf{v}_2, \mathbf{v}_3, \mathbf{v_m}\right)^T. \end{cases}$$

Finally, let $h: \{0,1\}^* \to \{0,1\}^\lambda$ be a collision-resistant hash function, the protocol is described as follows.

1. $\mathcal{P}$ samples

$$\begin{cases} \mathbf{Q}_0 \leftarrow \mathrm{GL}(m_0, q), \mathbf{Q} \leftarrow \mathrm{GL}(m, q); \mathbf{P}_1, \mathbf{P}_2, \mathbf{P}_3 \leftarrow \mathrm{GL}(n, q); \mathbf{P}_0 \leftarrow \mathrm{GL}(n_0, q); \\ \mathbf{v}_1, \mathbf{v}_2, \mathbf{v}_3 \leftarrow \mathbb{F}_{q^m}^n, \mathbf{v_m} \leftarrow S; \mathbf{v_x} \leftarrow \mathbb{F}_q^{N-1}, \mathbf{v_s} \leftarrow \mathbb{F}_{q^{m_0}}^{n_0}, \mathbf{v} \leftarrow \mathbb{F}_q^\ell; \\ \rho_1, \rho_2, \rho_3 \leftarrow 1^\lambda, \end{cases}$$

and sends the commitment $\mathsf{CMT} = (c_1, c_2, c_3)$, where

$$\begin{cases} c_1 = h\left(\mathbf{v}, \mathbf{Q}, \mathbf{Q}_0, \mathbf{P}_0, \ldots, \mathbf{P}_3, \mathbf{H} \cdot \mathbf{v_s}^T - \mathbf{A} \cdot \mathbf{v_x}^T, \widetilde{\mathbf{H}} \cdot \left(\mathbf{v}_1, \mathbf{v}_2, \mathbf{v}_3, \mathbf{v_m}\right)^T, \rho_1\right), \\ c_2 = h\left(\mathbf{Q}_0 \star \mathbf{v_s} \mathbf{P}_0, P_{\mathbf{v}}(\mathbf{v_x}), P'_{\mathbf{v}}(\mathbf{v_m}), (\mathbf{Q} \star \mathbf{v}_i \mathbf{P}_i)_{i=1}^3, \rho_2\right), \\ c_3 = h\left(\mathbf{Q}_0 \star (\mathbf{s} + \mathbf{v_s}) \mathbf{P}_0, P_{\mathbf{v}}(\mathbf{x} + \mathbf{v_x}), P'_{\mathbf{v}}(\mathbf{m} + \mathbf{v_m}), (\mathbf{Q} \star (\mathbf{r}_i + \mathbf{v}_i) \mathbf{P}_i)_{i=1}^3, \rho_3\right). \end{cases}$$

2. $\mathcal{V}$ sends a random challenge $\mathsf{Ch} \in \{1, 2, 3\}$ to $\mathcal{P}$.
3. $\mathcal{P}$ replies as:

- If $\mathsf{Ch} = 1$, reveal c_2 and c_3. Let $\mathbf{v}' = f_1(j) \cdot \mathbf{v}$.

$$\begin{cases} \widehat{\mathbf{v}}_{\mathbf{s}} = \mathbf{Q}_0 \star \mathbf{v}_{\mathbf{s}} \mathbf{P}_0, \\ \widehat{\mathbf{s}} = \mathbf{Q}_0 \star \mathbf{s} \mathbf{P}_0 \end{cases} \quad \widehat{\mathbf{v}}_{\mathbf{x}} = P_{\mathbf{v}}(\mathbf{v}_{\mathbf{x}}), \quad \widehat{\mathbf{v}}_{\mathbf{m}} = P'_{\mathbf{v}}(\mathbf{v}_{\mathbf{m}}), \quad \begin{cases} \widehat{\mathbf{v}}_i = \mathbf{Q} \star \mathbf{v}_i \mathbf{P}_i, \\ \widehat{\mathbf{r}}_i = \mathbf{Q} \star \mathbf{r}_i \mathbf{P}_i, \end{cases}$$

  $\mathcal{P}$ sends $\mathsf{RSP} = \big(\mathbf{v}', \widehat{\mathbf{s}}, \widehat{\mathbf{v}}_{\mathbf{s}}, \widehat{\mathbf{v}}_{\mathbf{x}}, \widehat{\mathbf{v}}_{\mathbf{m}}, (\widehat{\mathbf{v}}_i)_{i=1}^3, (\widehat{\mathbf{r}}_i)_{i=1}^3, \rho_2, \rho_3\big)$ to $\mathcal{V}$.

- If $\mathsf{Ch} = 2$, reveal c_1 and c_3. Let

$$\begin{cases} \mathbf{v}'' = \mathbf{v}, \mathbf{E} = \mathbf{Q}, \mathbf{E}_0 = \mathbf{Q}_0, \mathbf{F}_i = \mathbf{P}_i, 0 \leq i \leq 3, \\ \mathbf{z}_{\mathbf{s}} = \mathbf{s} + \mathbf{v}_{\mathbf{s}}, \mathbf{z}_{\mathbf{x}} = \mathbf{x} + \mathbf{v}_{\mathbf{x}}, \mathbf{z}_{\mathbf{m}} = \mathbf{m} + \mathbf{v}_{\mathbf{m}}, \mathbf{z}_i = \mathbf{r}_i + \mathbf{v}_i, 1 \leq i \leq 3. \end{cases}$$

  $\mathcal{P}$ sends $\mathsf{RSP} = \big(\mathbf{v}'', \mathbf{E}, \mathbf{E}_0, (\mathbf{F}_i)_{i=0}^3, \mathbf{z}_{\mathbf{s}}, \mathbf{z}_{\mathbf{x}}, \mathbf{z}_{\mathbf{m}}, (\mathbf{z}_i)_{i=1}^3, \rho_1, \rho_3\big)$ to $\mathcal{V}$.

- If $\mathsf{Ch} = 3$, reveal c_1 and c_2. Let

$$\begin{cases} \mathbf{v}''' = \mathbf{v}, \mathbf{U} = \mathbf{Q}, \mathbf{U}_0 = \mathbf{Q}_0, \mathbf{V}_i = \mathbf{P}_i, 0 \leq i \leq 3, \\ \mathbf{y}_{\mathbf{s}} = \mathbf{v}_{\mathbf{s}}, \mathbf{y}_{\mathbf{x}} = \mathbf{v}_{\mathbf{x}}, \mathbf{y}_{\mathbf{m}} = \mathbf{v}_{\mathbf{m}}, \mathbf{y}_i = \mathbf{v}_i, 1 \leq i \leq 3. \end{cases}$$

  $\mathcal{P}$ sends $\mathsf{RSP} = \big(\mathbf{v}''', \mathbf{U}, \mathbf{U}_0, (\mathbf{V}_i)_{i=0}^3, \mathbf{y}_{\mathbf{s}}, \mathbf{y}_{\mathbf{x}}, \mathbf{y}_{\mathbf{m}}, (\mathbf{y}_i)_{i=1}^3, \rho_1, \rho_2\big)$ to $\mathcal{V}$.

4. $\mathcal{V}$ performs the following checks:

- If $\mathsf{Ch} = 1$, let $\mathbf{w}_{\mathbf{x}} = \delta_{f_1^{-1}(\mathbf{v}')} \in \mathbb{F}_q^{N-1}$ and $\mathbf{w}_{\mathbf{m}} = f_2(\mathbf{v}') \in \mathbb{F}_{q^m}^k$. Check that $\widehat{\mathbf{s}} \in \mathcal{S}_{w_s}^{n_0, m_0}$ and $\widehat{\mathbf{r}}_i \in \mathcal{S}_{w_r}^{n, m}$ have the same support, and that

$$\begin{cases} c_2 = h\big(\widehat{\mathbf{s}}, \widehat{\mathbf{v}}_{\mathbf{x}}, \widehat{\mathbf{v}}_1, \widehat{\mathbf{v}}_2, \widehat{\mathbf{v}}_3, \widehat{\mathbf{v}}_{\mathbf{m}}, \rho_2\big), \\ c_3 = h\big(\widehat{\mathbf{s}} + \widehat{\mathbf{v}}_{\mathbf{s}}, \widehat{\mathbf{v}}_{\mathbf{x}} + \mathbf{w}_{\mathbf{x}}, \widehat{\mathbf{v}}_{\mathbf{m}} + \mathbf{w}_{\mathbf{m}}, (\widehat{\mathbf{r}}_i + \widehat{\mathbf{v}}_i)_{i=1}^3, \rho_3\big). \end{cases}$$

- If $\mathsf{Ch} = 2$, check that

$$\begin{cases} c_1 = h\left(\mathbf{v}'', \mathbf{E}, \mathbf{E}_0, \mathbf{F}_0, \dots, \mathbf{F}_3, \mathbf{H} \cdot \mathbf{z}_{\mathbf{s}}^T - \mathbf{A} \cdot \mathbf{z}_{\mathbf{x}}^T, \widetilde{\mathbf{H}} \cdot (\mathbf{z}_1, \mathbf{z}_2, \mathbf{z}_3, \mathbf{z}_{\mathbf{m}})^T - \mathbf{c}^T, \rho_1\right), \\ c_3 = h\big(\mathbf{E}_0 \star \mathbf{z}_{\mathbf{s}} \mathbf{F}_0, P_{\mathbf{v}''}(\mathbf{z}_{\mathbf{x}}), P'_{\mathbf{v}''}(\mathbf{z}_{\mathbf{m}}), (\mathbf{E} \star \mathbf{z}_i \mathbf{F}_i)_{i=1}^3, \rho_3\big). \end{cases}$$

- If $\mathsf{Ch} = 3$, check that

$$\begin{cases} c_1 = h\left(\mathbf{v}''', \mathbf{U}, \mathbf{U}_0, \mathbf{V}_0, \dots, \mathbf{V}_3, \mathbf{H} \cdot \mathbf{y}_{\mathbf{s}}^T - \mathbf{A} \cdot \mathbf{y}_{\mathbf{x}}^T, \widetilde{\mathbf{H}} \cdot (\mathbf{y}_1, \mathbf{y}_2, \mathbf{y}_3, \mathbf{y}_{\mathbf{m}})^T, \rho_1\right), \\ c_2 = h\big(\mathbf{U}_0 \star \mathbf{y}_{\mathbf{s}} \mathbf{V}_0, P_{\mathbf{v}'''}(\mathbf{y}_{\mathbf{x}}), P'_{\mathbf{v}'''}(\mathbf{y}_{\mathbf{m}}), (\mathbf{U} \star \mathbf{y}_i \mathbf{V}_i)_{i=1}^3, \rho_2\big). \end{cases}$$

5. $\mathcal{V}$ outputs 1 it all checks are passed; otherwise, it outputs 0.

3.2   Analysis

Proposition 1. *The above interactive protocol has perfect completeness, and has communication cost bounded by $C = (\ell + N - 1 + m_0^2 + m_0 n_0 + n_0^2 + m^2 + 3mn + 3n^2 + km) \log q + 5\lambda$. It is a statistical zero-knowledge argument in the random oracle model.*

Communication Cost

- The commitment CMT has bit-size 3λ.
- For Ch $= 1$, we have $C_1 = (\ell + 2n_0 m_0 + N - 1 + km + 6nm) \log q + 2\lambda$.
- For Ch $= 2$ or 3, we have

$$C_{2,3} = (\ell + N - 1 + m_0^2 + m_0 n_0 + n_0^2 + m^2 + 3mn + 3n^2 + km) \log q + 2\lambda.$$

The total cost is bounded by

$$C = (\ell + N - 1 + m_0^2 + m_0 n_0 + n_0^2 + m^2 + 3mn + 3n^2 + km) \log q + 5\lambda.$$

Zero-Knowledge Property

Lemma 2. *In the random oracle model, there exists an efficient simulator $\mathcal{S}$ interacting with a verifier $\widehat{\mathcal{V}}$, such that, given only the public input of the protocol, $\mathcal{S}$ outputs with probability negligibly close to $\frac{2}{3}$ a simulated transcript that is statistically close to the one produced by the honest prover in the real interaction.*

Proof. Simulator $\mathcal{S}$, given the public input $(\mathbf{H}, \mathbf{H}_1, \dots, \mathbf{H}_4, \mathbf{A}, \mathbf{c})$, starts by picking a random $\overline{\mathsf{Ch}} \in \{1, 2, 3\}$. Next, we consider 3 cases.

Case 1: $\overline{\mathsf{Ch}} = 1$, $\mathcal{S}$ proceeds as follows:

1. Compute $\mathbf{s}' \in \mathbb{F}_q^{n_0}$ and $\mathbf{x}' \in \mathbb{F}_q^{N-1}$ satisfying $\mathbf{H} \cdot \mathbf{s}'^T = \mathbf{A} \cdot \mathbf{x}'^T$, and $\mathbf{m}' \in S, \mathbf{r}_1', \mathbf{r}_2', \mathbf{r}_3' \in \mathbb{F}_q^m$ such that $\widetilde{\mathbf{H}} \cdot (\mathbf{r}_1', \mathbf{r}_2', \mathbf{r}_3', \mathbf{m}')^T = \mathbf{c}^T$.
2. Sample random objects, compute and send a commitment as in the real scheme. Namely, $\mathcal{S}$ samples

$$\begin{cases} \mathbf{Q}_0 \leftarrow \mathsf{GL}(m_0, q), \mathbf{Q} \leftarrow \mathsf{GL}(m, q); \mathbf{P}_1, \mathbf{P}_2, \mathbf{P}_3 \leftarrow \mathsf{GL}(n, q); \mathbf{P}_0 \leftarrow \mathsf{GL}(n_0, q); \\ \mathbf{v}_1, \mathbf{v}_2, \mathbf{v}_3 \leftarrow \mathbb{F}_q^m, \mathbf{v_m} \leftarrow S; \mathbf{v_x} \leftarrow \mathbb{F}_q^{N-1}, \mathbf{v_s} \leftarrow \mathbb{F}_q^{n_0}, \mathbf{v} \leftarrow \mathbb{F}_q^\ell; \\ \rho_1, \rho_2, \rho_3 \leftarrow 1^\lambda, \end{cases}$$

and sends the commitment $\mathsf{CMT} = (c_1', c_2', c_3')$, where

$$\begin{cases} c_1 = h(\mathbf{v}, \mathbf{Q}, \mathbf{Q}_0, \mathbf{P}_0, \dots, \mathbf{P}_3, \mathbf{H} \cdot \mathbf{v_s}^T - \mathbf{A} \cdot \mathbf{v_x}^T, \widetilde{\mathbf{H}} \cdot (\mathbf{v}_1, \mathbf{v}_2, \mathbf{v}_3, \mathbf{v_m})^T, \rho_1), \\ c_2 = h(\mathbf{Q}_0 \star \mathbf{v_s}\mathbf{P}_0, P_\mathbf{v}(\mathbf{v_x}), P_\mathbf{v}'(\mathbf{v_m}), (\mathbf{Q} \star \mathbf{v}_i \mathbf{P}_i)_{i=1}^3, \rho_2), \\ c_3 = h(\mathbf{Q}_0 \star (\mathbf{s} + \mathbf{v_s})\mathbf{P}_0, P_\mathbf{v}(\mathbf{x} + \mathbf{v_x}), P_\mathbf{v}'(\mathbf{m} + \mathbf{v_m}), (\mathbf{Q} \star (\mathbf{r}_i + \mathbf{v}_i)\mathbf{P}_i)_{i=1}^3, \rho_3). \end{cases}$$

Receiving a challenge Ch from $\widehat{\mathcal{V}}$, the simulator responds as follows:

- If Ch $= 1$: Output $\perp$, and abort.
- If Ch $= 2$: Send

$$\mathsf{RSP} = (\mathbf{v}, \mathbf{Q}, \mathbf{Q}_0, \mathbf{P}_1, \mathbf{P}_2, \mathbf{P}_3, \mathbf{s}' + \mathbf{v_s}, \mathbf{x}' + \mathbf{v_x}, \mathbf{m}' + \mathbf{v_m}, (\mathbf{r}_i' + \mathbf{v}_i)_{i=1}^3; \rho_1, \rho_3).$$

- If Ch $= 3$: Send

$$\mathsf{RSP} = (\mathbf{v}, \mathbf{Q}, \mathbf{Q}_0, \mathbf{P}_1, \mathbf{P}_2, \mathbf{P}_3, \mathbf{v_s}, \mathbf{v_x}, \mathbf{v_m}, \mathbf{v}_1, \mathbf{v}_2, \mathbf{v}_3; \rho_1, \rho_2).$$

Case 2: $\overline{\mathsf{Ch}} = 2$, $\mathcal{S}$ samples

$$\begin{cases} j' \leftarrow [N-1], \mathbf{s}' \leftarrow \mathcal{S}_{w_s}^{n_0, m_0}, \mathbf{r}'_1, \mathbf{r}'_2, \mathbf{r}'_3 \in \mathcal{S}_{w_r}^{n,m}; \\ \mathbf{Q}_0 \leftarrow \mathrm{GL}(m_0, q), \mathbf{Q} \leftarrow \mathrm{GL}(m, q); \mathbf{P}_1, \mathbf{P}_2, \mathbf{P}_3 \leftarrow \mathrm{GL}(n, q); \mathbf{P}_0 \leftarrow \mathrm{GL}(n_0, q); \\ \mathbf{v}_1, \mathbf{v}_2, \mathbf{v}_3 \leftarrow \mathbb{F}_{q^m}^n, \mathbf{v_m} \leftarrow S; \mathbf{v_x} \leftarrow \mathbb{F}_q^{N-1}, \mathbf{v_s} \leftarrow \mathbb{F}_{q^{m_0}}^{n_0}, \mathbf{v} \leftarrow \mathbb{F}_q^\ell; \\ \rho_1, \rho_2, \rho_3 \leftarrow 1^\lambda, \end{cases}$$

and let $\mathbf{x}' = \delta_{j'}^{N-1}$, and $\mathbf{m}' = \mathsf{I2V}(j')$. $\mathcal{S}$ sends the commitment computed as in the case $\overline{\mathsf{Ch}} = 1$. After receiving a challenge Ch from $\widehat{\mathcal{V}}$, it responds as follows:

- If $\mathsf{Ch} =: 1$ Send

  $$\mathsf{RSP} = \big(\mathsf{I2V}(j') + \mathbf{v}, \mathbf{Q}_0 \star \mathbf{sP}_0, \mathbf{Q}_0 \star \mathbf{v_s P}_0, P_\mathbf{v}(\mathbf{v_x}), P'_\mathbf{v}(\mathbf{v_m}), \mathbf{Q} \star \mathbf{v}_i \mathbf{P}_i, \mathbf{Q} \star \mathbf{r}'_i \mathbf{P}_i; \rho_2, \rho_3\big),$$

  which contains all $i = 1, 2, 3$.
- If $\mathsf{Ch} = 2$: Output $\perp$, and abort.
- If $\mathsf{Ch} = 3$: Send RSP computed as in the case $(\overline{\mathsf{Ch}} = 1, \mathsf{Ch} = 3)$.

Case 3: $\overline{\mathsf{Ch}} = 3$, the simulator performs the preparation as in the case $\overline{\mathsf{Ch}} = 2$. It sends the commitment $\mathsf{CMT} = (c'_1, c'_2, c'_3)$, where c'_2 and c'_3 are computed as usual, while

$$c'_1 = h\left(\mathbf{v}, \mathbf{Q}, \mathbf{Q}_0, \mathbf{P}_0, \dots, \mathbf{P}_3, \mathbf{H} \cdot (\mathbf{s}' + \mathbf{v_s})^T - \mathbf{A} \cdot (\mathbf{x}' + \mathbf{v_x})^T, \widetilde{\mathbf{H}} \cdot \mathbf{z}^T, \rho_1\right),$$

where $\mathbf{z} = (\mathbf{r}'_1 + \mathbf{v}_1, \mathbf{r}'_2 + \mathbf{v}_2, \mathbf{r}'_3 + \mathbf{v}_3, \mathbf{m}' + \mathbf{v_m})$. Next, after receiving a challenge Ch, $\mathcal{S}$ responds as follows:

- If $\mathsf{Ch} = 1$: Send RSP as in the case $(\overline{\mathsf{Ch}} = 2, \mathsf{Ch} = 1)$.
- If $\mathsf{Ch} = 2$: Send RSP as in the case $(\overline{\mathsf{Ch}} = 1, \mathsf{Ch} = 2)$.
- If $\mathsf{Ch} = 3$: Output $\perp$, and abort.

Since the challenge is a random value from $\{1, 2, 3\}$, so the probability that $\mathcal{S}$ outputs $\perp$ is $\frac{1}{3}$. In the cases that $\mathcal{S}$ does not output $\perp$, one can easily verify that the distribution of its output is identical to that in the real interaction. $\square$

Soundness Property

Lemma 3. *Given the public input of the protocol, a commitment* CMT *and 3 valid responses* $\mathsf{RSP}_1, \mathsf{RSP}_2, \mathsf{RSP}_3$ *to all possible values of the challenge* Ch, *one can efficiently construct a knowledge extractor* $\mathcal{E}$ *that outputs a tuple*

$$(j', \mathbf{s}', \mathbf{r}'_1, \mathbf{r}'_2, \mathbf{r}'_3) \in [N-1] \times \mathbb{F}_{q^{m_0}}^{n_0} \times \mathbb{F}_{q^m}^n \times \mathbb{F}_{q^m}^n \times \mathbb{F}_{q^m}^n$$

such that

$$\begin{cases} \mathbf{H} \cdot \mathbf{s}'^T = \mathbf{y}'^T & and \quad \mathbf{s}' \in \mathcal{S}_{w_s}^{n_0, m_0}, \\ \widetilde{\mathbf{H}} \cdot \big(\mathbf{r}'_1, \mathbf{r}'_2, \mathbf{r}'_3, \mathsf{I2V}(j')\big)^T = \mathbf{c}^T & and \quad \mathbf{r}'_i \in \mathcal{S}_{w_r}^{n,m}, i = 1, 2, 3. \end{cases}$$

Proof. Assume that we have a commitment $\mathsf{CMT} = (c_1, c_2, c_3)$ and 3 responses

$$\begin{cases} \mathsf{RSP}_1 = \left(\mathbf{v}', \widehat{\mathbf{s}}, \widehat{\mathbf{v}}_\mathbf{s}, \widehat{\mathbf{v}}_\mathbf{x}, \widehat{\mathbf{v}}_\mathbf{m}, (\widehat{\mathbf{v}}_i)_{i=1}^3, (\widehat{\mathbf{r}}_i)_{i=1}^3, \rho_2, \rho_3\right), \\ \mathsf{RSP}_2 = \left(\mathbf{v}'', \mathbf{E}, \mathbf{E}_0, (\mathbf{F}_i)_{i=0}^3, \mathbf{z}_\mathbf{s}, \mathbf{z}_\mathbf{x}, \mathbf{z}_\mathbf{m}, (\mathbf{z}_i)_{i=1}^3, \rho_1, \rho_3\right), \\ \mathsf{RSP}_3 = \left(\mathbf{v}''', \mathbf{U}, \mathbf{U}_0, (\mathbf{V}_i)_{i=0}^3, \mathbf{y}_\mathbf{s}, \mathbf{y}_\mathbf{x}, \mathbf{y}_\mathbf{m}, (\mathbf{y}_i)_{i=1}^3, \rho_1, \rho_2\right) \end{cases}$$

that satisfy all the verification conditions with respect to $\mathsf{Ch} = 1, 2, 3$, respectively. Thus, we have the following relations:

$$\begin{cases} \widehat{\mathbf{s}} \in \mathcal{S}_{w_s}^{n_0, m_0}, \mathbf{w}_\mathbf{x} = \delta_{f_1^{-1}(\mathbf{v}')}^{N-1}, \mathbf{w}_\mathbf{m} = f_2(\mathbf{v}'), \widehat{\mathbf{r}}_i \in \mathcal{S}_{w_r}^{n, m}, \\ c_1 = h\left(\mathbf{v}'', \mathbf{E}, \mathbf{E}_0, \mathbf{F}_0, \ldots, \mathbf{F}_3, \mathbf{H} \cdot \mathbf{z}_\mathbf{s}^T - \mathbf{A} \cdot \mathbf{z}_\mathbf{x}^T, \widetilde{\mathbf{H}} \cdot (\mathbf{z}_1, \mathbf{z}_2, \mathbf{z}_3, \mathbf{z}_\mathbf{m})^T - \mathbf{c}^T, \rho_1\right), \\ c_1 = h\left(\mathbf{v}''', \mathbf{U}, \mathbf{U}_0, \mathbf{V}_0, \ldots, \mathbf{V}_3, \mathbf{H} \cdot \mathbf{y}_\mathbf{s}^T - \mathbf{A} \cdot \mathbf{y}_\mathbf{x}^T, \widetilde{\mathbf{H}} \cdot (\mathbf{y}_1, \mathbf{y}_2, \mathbf{y}_3, \mathbf{y}_\mathbf{m})^T, \rho_1\right), \\ c_2 = h\left(\widehat{\mathbf{s}}, \widehat{\mathbf{v}}_\mathbf{x}, \widehat{\mathbf{v}}_1, \widehat{\mathbf{v}}_2, \widehat{\mathbf{v}}_3, \widehat{\mathbf{v}}_\mathbf{m}, \rho_2\right), \\ c_2 = h\left(\mathbf{U}_0 \star \mathbf{y}_\mathbf{s} \mathbf{V}_0, P_{\mathbf{v}'''}(\mathbf{y}_\mathbf{x}), P'_{\mathbf{v}'''}(\mathbf{y}_\mathbf{m}), (\mathbf{U} \star \mathbf{y}_i \mathbf{V}_i)_{i=1}^3, \rho_2\right), \\ c_3 = h\left(\widehat{\mathbf{s}} + \widehat{\mathbf{v}}_\mathbf{s}, \widehat{\mathbf{v}}_\mathbf{x} + \mathbf{w}_\mathbf{x}, \widehat{\mathbf{v}}_\mathbf{m} + \mathbf{w}_\mathbf{m}, (\widehat{\mathbf{r}}_i + \widehat{\mathbf{v}}_i)_{i=1}^3, \rho_3\right), \\ c_3 = h\left(\mathbf{E}_0 \star \mathbf{z}_\mathbf{s} \mathbf{F}_0, P_{\mathbf{v}''}(\mathbf{z}_\mathbf{x}), P'_{\mathbf{v}''}(\mathbf{z}_\mathbf{m}), (\mathbf{E} \star \mathbf{z}_i \mathbf{F}_i)_{i=1}^4, \rho_3\right). \end{cases}$$

Since h is a collision-resistant hash function, it must be that:

$$\begin{cases} \mathbf{v}'' = \mathbf{v}''', \mathbf{E} = \mathbf{U}, \mathbf{E}_0 = \mathbf{U}_0, \mathbf{F}_i = \mathbf{V}_i, \\ \delta_{f_1^{-1}(\mathbf{v}')}^{N-1} = \mathbf{w}_\mathbf{x} = P_{\mathbf{v}''}(\mathbf{z}_\mathbf{x}) - P_{\mathbf{v}'''}(\mathbf{y}_\mathbf{x}) = P_{\mathbf{v}''}(\mathbf{z}_\mathbf{x} - \mathbf{y}_\mathbf{x}), \\ f_2(\mathbf{v}') = \mathbf{w}_\mathbf{m} = P'_{\mathbf{v}''}(\mathbf{z}_\mathbf{m}) - P'_{\mathbf{v}'''}(\mathbf{y}_\mathbf{m}) = P'_{\mathbf{v}''}(\mathbf{z}_\mathbf{m} - \mathbf{y}_\mathbf{m}), \\ \widehat{\mathbf{s}} = \mathbf{E}_0 \star \mathbf{z}_\mathbf{s} \mathbf{F}_0 - \mathbf{U}_0 \star \mathbf{y}_\mathbf{s} \mathbf{V}_0 = \mathbf{E}_0 \star (\mathbf{z}_\mathbf{s} - \mathbf{y}_\mathbf{s}) \mathbf{F}_0 \in \mathcal{S}_{w_s}^{n_0, m_0}, \\ \widehat{\mathbf{r}}_i = \mathbf{E} \star \mathbf{z}_i \mathbf{F}_i - \mathbf{U} \star \mathbf{y}_i \mathbf{V}_i = \mathbf{E} \star (\mathbf{z}_i - \mathbf{y}_i) \mathbf{F}_i \in \mathcal{S}_{w_r}^{n, m}, \\ \mathbf{H} \cdot (\mathbf{z}_\mathbf{s} - \mathbf{y}_\mathbf{s})^T - \mathbf{A} \cdot (\mathbf{z}_\mathbf{x} - \mathbf{y}_\mathbf{x})^T = \mathbf{0}, \\ \widetilde{\mathbf{H}} \cdot (\mathbf{z}_1 - \mathbf{y}_1, \mathbf{z}_2 - \mathbf{y}_2, \mathbf{z}_3 - \mathbf{y}_3, \mathbf{z}_\mathbf{m} - \mathbf{y}_\mathbf{m})^T = \mathbf{c}^T. \end{cases}$$

Now, let

- $j' = f_1^{-1}(\mathbf{z}_\mathbf{x} - \mathbf{y}_\mathbf{x}) \in [N-1]$,
- $\mathbf{s}' = \mathbf{z}_\mathbf{s} - \mathbf{y}_\mathbf{s} \in \mathcal{S}_{w_s}^{n_0, m_0}$,
- $\mathbf{r}'_i = \mathbf{z}_i - \mathbf{y}_i \in \mathcal{S}_{w_r}^{n, m}$,

It is easy to see that they satisfy the lemma. $\square$

4   Our Code-Based Group Signature Scheme

1. **KeyGen**$(1^\lambda, N-1)$: On input a security parameter λ and an expected number of group users $N - 1 = q^\ell - 1$, the algorithm first prepares as follows:
   - A primitive element α of $\mathbb{F}_q$ to describe the map f_1.
   - Parameters $m = m(\lambda), n = n(\lambda), k = k(\lambda), t = t(\lambda)$ for a rank Gabidulin code $[n, k, 2t+1]$ over $\mathbb{F}_{q^m}$. In addition, it chooses an irreducible polynomial $F(X) \in \mathbb{F}_q[X]$ of degree m to define $\mathbb{F}_{q^m}$ as $\mathbb{F}_q[X]/\langle F \rangle$, and an irreducible polynomial $F_0(X) \in \mathbb{F}_q[X]$ of degree m_0 to define $\mathbb{F}_{q^{m_0}}$ as $\mathbb{F}_q[X]/\langle F_0 \rangle$.

- Parameters $w_r = w_r(\lambda)$ and an irreducible polynomial $P(X) \in \mathbb{F}_q[X]$ of degree n such that $P(X)$ is also irreducible over $\mathbb{F}_{q^m}$.
- Parameters $n_0 = n_0(\lambda), r_0 = r_0(\lambda), w_s = w_s(\lambda)$ for the syndrome decoding problem. We choose $r_0 = \frac{1}{2}n_0$.
- An irreducible polynomial $p(X)$ of degree ℓ over $\mathbb{F}_q$.
- A generator matrix $\mathbf{M} \in \mathbb{F}_q^{\ell \times mk}$ of systematic form of a public linear code $\mathcal{C}$ over $\mathbb{F}_q$.
- Two collision-resistant hash functions h and $\mathcal{H}$ used for generating commitments and random challenges, respectively.

Then the algorithm performs the following steps:

1. Run $\mathsf{RQC}(m, n, k, w_r, P(X))$ for a key pair $(\mathsf{pk}_{\mathsf{RQC}} = (\mathbf{H}_1, \dots, \mathbf{H}_4), \mathsf{sk}_{\mathsf{RQC}})$.
2. The matrix $\mathbf{H}$ is constructed in the following way: Choose an irreducible polynomial $P_0(X) \in \mathbb{F}_q[X]$ of degree r_0 so that it is also irreducible over $\mathbb{F}_{q^{m_0}}$, a random vector $\mathbf{h}_0 \in \mathbb{F}_{q^{m_0}}^{r_0}$; then $\mathbf{H}$ is the ideal matrix generated by $(\mathbf{h}_0, P_0(X))$.
3. For each $j \in [N-1]$, choose $\mathbf{s}_j \leftarrow \mathcal{S}_{w_s}^{n_0, m_0}$ and let $\mathbf{y}_j^T = \mathbf{H} \cdot \mathbf{s}_j^T$, set $\mathbf{A} = [\mathbf{y}_1^T | \cdots | \mathbf{y}_{N-1}^T]$.
4. Set $\widetilde{\mathbf{H}} = [\mathbf{H}_1 | \cdots | \mathbf{H}_4]$ and output

$$\mathsf{gpk} = (\mathbf{H}, \widetilde{\mathbf{H}}, \mathbf{A}, p, P, P_0, F, F_0, \mathbf{M}, \alpha), \mathsf{gmsk} = \mathsf{sk}_{\mathsf{RQC}}, \mathsf{gsk} = (\mathbf{s}_1, \dots, \mathbf{s}_{N-1}).$$

2. **Sign**$(\mathsf{gsk}[j], M)$: To sign a message $M \in \{0, 1\}^*$ under gpk, the group user of Index j performs the following steps:

- Encrypt the representation vector of j, i.e., the vector $\mathsf{I2V}(j) \in \mathbb{F}_{q^m}^k$, using $\mathsf{pk}_{\mathsf{RQC}}$.
- Generate an NIZKAoK Π to simultaneously prove the possession of a vector $\mathbf{s} \in \mathcal{S}_{w_s}^{n_0, m_0}$ corresponding to a certain syndrome $\mathbf{y} \in \{\mathbf{y}_1, \dots, \mathbf{y}_{N-1}\}$ with hidden index j, and that $\mathbf{c} = (\mathbf{c}_1, \mathbf{c}_2)$ is a correct RQC encryption of $\mathsf{I2V}(j)$. This is done by using the interactive protocol in the above section with public input $(\mathbf{H}, \mathbf{H}_1, \dots, \mathbf{H}_4, \mathbf{A}, \mathbf{c})$, and prover's witness $(j, \mathbf{s}, \mathbf{r}_1, \mathbf{r}_2, \mathbf{r}_3)$ that satisfies

$$\begin{cases} \mathbf{H} \cdot \mathbf{s}^T = \mathbf{y}_j^T \quad \text{and} \quad \mathbf{s} \in \mathcal{S}_{w_s}^{n_0, m_0}, \\ [\mathbf{H}_1 | \cdots | \mathbf{H}_4] \cdot (\mathbf{r}_1, \mathbf{r}_2, \mathbf{r}_3, \mathsf{I2V}(j))^T = (\mathbf{c}_1, \mathbf{c}_2)^T. \end{cases}$$

The protocol is repeated $\kappa = \omega(\log \lambda)$ times to achieve negligible soundness error, and then made non-interactive, i.e., we have

$$\Pi = (\mathsf{CMT}^1, \dots, \mathsf{CMT}^\kappa; (\mathsf{Ch}^1, \dots, \mathsf{Ch}^\kappa); \mathsf{RSP}^1, \dots, \mathsf{RSP}^\kappa),$$

where $(\mathsf{Ch}^1, \dots, \mathsf{Ch}^\kappa) = \mathcal{H}(M; \mathsf{CMT}^1, \dots, \mathsf{CMT}^\kappa; \mathsf{gpk}, \mathbf{c})$.

- Output the group signature $\Sigma = (\mathbf{c}, \Pi)$.

3. **Verify**$(\mathsf{gpk}, M, \Sigma)$: Parse Σ as $(\mathbf{c}, \Pi)$, and parse Π as above, and proceed as follows:

- If $(\mathsf{Ch}^1, \dots, \mathsf{Ch}^\kappa) \neq \mathcal{H}(M; \mathsf{CMT}^1, \dots, \mathsf{CMT}^\kappa; \mathsf{gpk}, \mathbf{c})$, then return 0.

- For $i = 1$ to κ, run the verification step of the interactive protocol with public input $(\mathbf{H}, \mathbf{H}_1, \ldots, \mathbf{H}_4, \mathbf{A}, \mathbf{c})$ to check the validity of RSP^i with respect to CMT^i and Ch^i. If any of the verification does not hold true, then return 0.
- Return 1.

4. **Open**(gmsk, M, Σ): Parse Σ as $(\mathbf{c}, \Pi)$ and run $\mathsf{RQC.Dec}(\mathsf{gmsk}, \mathbf{c})$ to decrypt $\mathbf{c}$. If the decryption fails, then return $\perp$. If the decryption outputs $\mathbf{v} \in \mathbb{F}_{q^m}^k$, then return $j = \mathsf{V2I}(\mathbf{v}) \in [N-1]$.

The security of the scheme is stated in the following theorem.

Theorem 1. *In the random oracle model:*

- *If the decisional $2-\mathsf{IRSD}(n, w_r)$ and $3-\mathsf{IRSD}(n, w_r)$ problems are hard, then the scheme is* CPA- *anonymous.*
- *If the ideal rank syndrome decoding problem $2-\mathsf{IRSD}(n_0, w_s)$ is hard, then the scheme is traceable.*

4.1 Efficiency and Correctness

Efficiency. The size of gpk is dominated by $T \cdot \log q$, where

$$T = r_0 m_0 N + (k+2)nm + (\ell + n + r_0 + m_0 + m) + \ell k m.$$

The length of the NIZKAoK is κ times the communication cost of the underlying interactive protocol. Therefore, the size of Σ is bounded by

$$\left((\ell + N - 1 + m_0^2 + m_0 n_0 + n_0^2 + m^2 + 3mn + 3n^2 + km)\log q + 5\lambda\right) \cdot \kappa + n.$$

Correctness. By guaranteeing that the user is honest and the underlying interactive protocol is perfectly complete, the correctness of the scheme is easily verified.

4.2 Anonymity

Let $\mathcal{A}$ be a PPT adversary attacking the CPA-anonymity of the scheme with advantage ε. We prove that ε is a negligible function of λ by considering the following sequence of experiments.

Experiment $G_0^{(b)}$. The challenger runs **KeyGen** to obtain

$$\begin{cases} \mathsf{gpk} = (\mathbf{H}, \widetilde{\mathbf{H}}, \mathbf{A}, p, P, P_0, F, F_0, \mathbf{M}, \alpha), \\ \mathsf{gmsk} = \mathsf{sk}_{\mathsf{RQC}}, \\ \mathsf{gsk} = (\mathbf{s}_1, \ldots, \mathbf{s}_{N-1}), \end{cases}$$

then gives gpk and gsk to $\mathcal{A}$. In the challenge phase, $\mathcal{A}$ outputs a message M^* and two indices $j_0, j_1 \in [N-1]$. The challenger sends back a challenge signature

$$\Sigma^* = (\mathbf{c}^*, \Pi^*) \leftarrow \mathbf{Sign}(\mathbf{s}_{j_b}, M^*).$$

The adversary outputs b with probability $\frac{1}{2} + \varepsilon$.

Experiment $G_1^{(b)}$. The challenge simulates Π^* as follows:

1. Compute $\mathbf{c}^*$ as in the Experiment $G_0^{(b)}$.
2. Run the simulator of the underlying interactive protocol and programming $\mathcal{H}$ accordingly.
3 Output the simulated of Π^*.

By the property of the simulator, we have $G_0^{(b)}$ and $G_1^{(b)}$ are statistically closed.

Experiment $G_2^{(b)}$. In this experiment, the vector $\mathbf{s}$ is replaced by a random vector in $\mathbb{F}_{q^m}^n$. By the hardness of the decisional 2-IRSD(n, w_r) problem, the adversary cannot distinguish a real public key $\mathbf{s}$ from a random one. Thus, Experiments $G_2^{(b)}$ and $G_1^{(b)}$ are computationally indistinguishable.

Experiment $G_3^{(b)}$. The vectors $\mathbf{r}_1, \mathbf{r}_2, \mathbf{r}_3$ are replaced by random vectors $\overline{\mathbf{r}}_1, \overline{\mathbf{r}}_2, \overline{\mathbf{r}}_3 \leftarrow \mathbb{F}_{q^m}^n$. By the hardness of the decisional $3 - $ IRSD(n, w_r) problem, we have $\mathbf{r}_1 + \mathbf{h} \cdot \mathbf{r}_2$ and $\mathbf{r}_3 + \mathbf{s} \cdot \mathbf{r}_2$ are computationally indistinguishable from $\overline{\mathbf{r}}_1 + \mathbf{h} \cdot \overline{\mathbf{r}}_2$ and $\overline{\mathbf{r}}_3 + \mathbf{s} \cdot \overline{\mathbf{r}}_2$, respectively. As a consequence, $\mathbf{r}_3 + \mathbf{s} \cdot \mathbf{r}_2 + \mathsf{I2V}(j_b) \cdot \mathbf{G}$ and $\overline{\mathbf{r}}_3 + \mathbf{s} \cdot \overline{\mathbf{r}}_2 + \mathsf{I2V}(j_b) \cdot \mathbf{G}$ are computationally indistinguishable. Therefore, Experiments $G_2^{(b)}$ and $G_3^{(b)}$ are computationally indistinguishable.

Experiment $\bar{G}_4$. In this experiment, the ciphertext is set as $\mathbf{c}^* \leftarrow \mathbb{F}_{q^m}^{2n}$. It is evident that the distribution of $\mathbf{c}^*$ in Experiments $G_3^{(b)}$ and G_4 are identical, and hence, $G_3^{(b)}$ and G_4 are statistically indistinguishable. Observe that the ciphertext now no longer depends on the challenger's bit b, therefore, $\mathcal{A}$'s advantage in this experiment is 0.

The above arguments show that the advantage of $\mathcal{A}$ in $G_0^{(b)}$ is negligible, *i.e.,* ε is negligible. Thus, the scheme is CPA-anonymous.

4.3   Traceability

The proof of this property is quite similar to that of [9]. The only difference is that our proof is for rank metric. We include it here for the sake of completeness.

Assume that $\mathcal{A}$ is a PPT traceability adversary against our group signature scheme with success probability ε. We construct an algorithm $\mathcal{F}$ that solves the RSD(n_0, r_0, w_s) problem with success probability polynomially related to ε.

At first, $\mathcal{F}$ receives a challenge from a decisional $2 - $ IRSD(n_0, w_s) instance, *i.e.,* a random pair $(\overline{\mathbf{H}}, \widetilde{\mathbf{y}}) \in \mathbb{F}_{q^{m_0}}^{r_0 \times n_0} \times \mathbb{F}_{q^{m_0}}^{r_0}$, where $\overline{\mathbf{H}}$ is the ideal matrix described by $(\mathbf{h}_0, P_0)$. The task of $\mathcal{F}$ is to find a vector $\mathbf{s} \in \mathcal{S}_{w_s}^{n_0, m_0}$ such that $\overline{\mathbf{H}} \cdot \mathbf{s}^T = \overline{\mathbf{y}}^T$. It then proceeds as follows:

1. Pick a guess j^* and set $\mathbf{y}_{j^*} = \widetilde{\mathbf{y}}$.
2. Set $\mathbf{H} = \overline{\mathbf{H}}$. For each $j \in [N-1] \backslash \{j^*\}$, sample $\mathbf{s}_j \leftarrow \mathcal{S}_{w_s}^{n_0, m_0}$ and set $\mathbf{y}_j = \mathbf{s}_j \cdot \mathbf{H}^T$.
3. Run RQC.KeyGen(n, k, w_r) to obtain a key pair $(\mathsf{pk}_{\mathsf{RQC}}, \mathsf{sk}_{\mathsf{RQC}})$.
4. Send $\mathsf{gpk} = (\mathbf{H}, \widetilde{\mathbf{H}}, \mathbf{A}, p, P, P_0, F, F_0, \mathbf{M}, \alpha)$ and $\mathsf{gmsk} = \mathsf{sk}_{\mathsf{RQC}}$ to $\mathcal{A}$.

Here, since the decisional $2 - \mathsf{IRSD}(n_0, w_s)$ is hard, so the view of $\mathcal{A}$ on the instance produced by $\mathcal{F}$ is computationally indistinguishable to its view on the instance from the real protocol. Next, $\mathcal{F}$ responds to the queries from $\mathcal{A}$. It initializes a set $CU = \varnothing$, and proceeds as follows:

1. For queries to the random oracle $\mathcal{H}$, it returns uniformly random values in $\{1, 2, 3\}^{\kappa}$. Suppose that $\mathcal{A}$ makes $Q_{\mathcal{H}}$ queries to the random oracle, then for each $\eta \leq Q_{\mathcal{H}}$, we let r_{η} be the answer to the η-th query.
2. For query to $\mathcal{O}^{\mathsf{Corrupt}}(j)$, if $j = j^*$, then $\mathcal{F}$ aborts; if $j \neq j^*$, then $\mathcal{F}$ sets $CU := CU \cup \{j\}$ and gives $\mathbf{s}_j$ to $\mathcal{A}$.
3. For query to $\mathcal{O}^{\mathsf{Sign}}(j, M)$, for $j \in [N - 1]$ and any message M:
   - If $j \neq j^*$, then $\mathcal{F}$ honestly computes a signature by using $\mathbf{s}_j$.
   - If $j = j^*$, then $\mathcal{F}$ returns a simulated signature Σ^*.

At some point, $\mathcal{A}$ outputs a forged signature Σ^* on some message M^*, where

$$\Sigma^* = \left(\mathbf{c}^*, \mathsf{CMT}^{(1)}, \ldots, \mathsf{CMT}^{(\kappa)}; \mathsf{Ch}^{(1)}, \ldots, \mathsf{Ch}^{(\kappa)}; \mathsf{RSP}^{(1)}, \ldots, \mathsf{RSP}^{(\kappa)}\right).$$

This signature must satisfy all the requirements of the traceability experiment. Now $\mathcal{F}$ uses $\mathsf{sk}_{\mathsf{RQC}}$ to open Σ^*. It aborts if the opening algorithm does not output j^*. The probability that $\mathcal{F}$ aborts is at most $\frac{N-1}{N} + \left(\frac{2}{3}\right)^{\kappa}$. Therefore, with probability at least $\frac{1}{N} - \left(\frac{2}{3}\right)^{\kappa}$, it holds that

$$\begin{cases} \mathsf{Verify}(\mathsf{gpk}, M^*, \Sigma^*) = 1, \\ \mathsf{Open}(\mathsf{sk}_{\mathsf{RQC}}, M^*, \Sigma^*) = j^*. \end{cases}$$

Assume that the above equalities hold, we denote Δ the tuple

$$\left(M^*; \mathsf{CMT}^{(1)}, \ldots, \mathsf{CMT}^{(\kappa)}; \mathbf{H}, \mathbf{H}_1, \ldots, \mathbf{H}_4, \mathbf{y}_1, \ldots, \mathbf{y}_{N-1}, \mathbf{c}^*\right).$$

Observe that with probability at least $p = \varepsilon - 3^{-\kappa}$, there exists a certain $\eta^* \leq Q_{\eta}$ such that Δ was the input of the η^*-th query. $\mathcal{F}$ picks η^* as the target forking point and replays $\mathcal{A}$ many times with the same random tape and input. In each run, for the first $\eta^* - 1$ queries, $\mathcal{A}$ is given the same answers $r_1, \ldots, r_{\eta^*-1}$ as in the initial run; from the η^*-th query onwards, $\mathcal{F}$ answers with fresh random values in $\{1, 2, 3\}^{\kappa}$. With probability greater than $\frac{1}{2}$ and within $32 \cdot Q_{\mathcal{H}}/p$ executions of $\mathcal{A}$, algorithm $\mathcal{F}$ can obtain a 3-fork, say

$$r_{1,\eta^*} = \left(\mathsf{Ch}_1^{(1)}, \ldots, \mathsf{Ch}_1^{(\kappa)}\right),$$
$$r_{2,\eta^*} = \left(\mathsf{Ch}_2^{(1)}, \ldots, \mathsf{Ch}_2^{(\kappa)}\right),$$
$$r_{3,\eta^*} = \left(\mathsf{Ch}_3^{(1)}, \ldots, \mathsf{Ch}_3^{(\kappa)}\right).$$

Note that one has

$$\Pr\left[i \in \{1, \ldots, \kappa\} | \{\mathsf{Ch}_1^{(i)}, \mathsf{Ch}_2^{(i)}, \mathsf{Ch}_3^{(i)}\} = \{1, 2, 3\}\right] = 1 - \left(\frac{7}{9}\right)^{\kappa}.$$

Conditioned on the existence of such index i, one parses the 3 forgeries corresponding to the fork to obtain $(\mathsf{RSP}_1^{(i)}, \mathsf{RSP}_2^{(i)}, \mathsf{RSP}_3^{(i)})$. They are three valid responses to three different challenges of the same commitment $\mathsf{CMT}^{(i)}$. By using the knowledge extractor as in Lemma 3, one can efficiently find a valid solution to the challenge $\mathsf{RSD}(n_0, r_0, w_s)$ instance $(\overline{\mathbf{H}}, \overline{\mathbf{y}})$.

Finally, if $\mathcal{A}$ has success probability ε and running time T in attacking the traceability of our group signature scheme, then $\mathcal{F}$ has success probability at least $\frac{1}{2} \cdot \left(\frac{1}{N} - \left(\frac{2}{3}\right)^\kappa\right) \cdot \left(1 - \left(\frac{7}{9}\right)^\kappa\right)$ and running time $32 \cdot T \cdot Q_\mathcal{H}/(\varepsilon - 3^{-\kappa}) + \mathrm{poly}(\lambda, N)$.

5   Parameters

In this section, we give a few examples of parameters for our code-based group signature scheme. The parameters are chosen so that the attacks in [5] and [10] have complexity at least at level 2^{128} to solve the $\mathsf{RSD}(n_0, r_0, w_s)$ problem or to break the RQC scheme.

- We consider $q = 2$, and thus $\log 2 = 1$. In this case, the map f_1 becomes the representation map with respect to the base 2.
  Parameters for the RQC scheme: $m = 139, n = 101, k = 5$ which are taken from [1].
- Parameter for the syndrome decoding problem corresponding to the matrix $\mathbf{H}$: $m_0 = 47, r_0 = 43, n_0 = 86, w_s = 7$.
- The number of users $N - 1 = q^\ell - 1$ for $\ell \in \{4, 8, \ldots, 24\}$.

We have the following table ($\kappa = 220$) (Table 2).

Table 2. Example of parameters.

ℓ	PK size	Signatures Size
4	16.71 KB	2.94 MB
8	77.68 KB	2.95 MB
12	1.05 MB	3.06 MB
16	16.57 MB	4.74 MB
20	264.9 MB	31.78 MB
24	4.24 GB	464.3 MB

6   Conclusion

In this work, we have constructed a code-based group signature scheme in the rank metric context. In some cases, our parameters are better than those of the previous work as in [9].

One feature of our scheme may be noteworthy, that is, in the second layer, we made use of RQC, however, one can use HQC due to the same structure as RQC. The only shortcoming is that the signature size would be larger.

References

1. Aguila, C., et al.: Rank quasi cyclic (RQC) first round submission to the NIST post-quantum cryptography call, November 2017
2. Aguilar, C., Blazy, O., Deneuville, J.-C., Gaborit, P., Zémor, G.: Efficient encryption from random quasi-cyclic codes. IEEE Trans. Inf. Theory **64**, 3927–3943 (2018)
3. Alamélou, Q., Blazy, O., Cauchie, S., and Gaborit, P.: A code-based group signature scheme. Presented at WCC, April 2015
4. Alamélou, Q., Blazy, O., Cauchie, S., Gaborit, P.: A practical group signature scheme based on rank metric. In: Duquesne, S., Petkova-Nikova, S. (eds.) WAIFI 2016. LNCS, vol. 10064, pp. 258–275. Springer, Cham (2016). https://doi.org/10.1007/978-3-319-55227-9_18
5. Aragon, N., Gaborit, P., Hauteville, A., Tillich, J.-P.: A new algorithm for solving the rank syndrome decoding problem. In: EEE International Symposium on Information Theory, ISIT 2018, Vail, CO, USA, 17–22 June 2018, pp. 2421–2425 (2018)
6. Becker, A., Joux, A., May, A., Meurer, A.: Decoding random binary linear codes in $2^{n/20}$: how $1+1=0$ improves information set decoding. In: Pointcheval, D., Johansson, T. (eds.) EUROCRYPT 2012. LNCS, vol. 7237, pp. 520–536. Springer, Heidelberg (2012). https://doi.org/10.1007/978-3-642-29011-4_31
7. Bellare, M., Micciancio, D., Warinschi, B.: Foundations of group signatures: formal definitions, simplified requirements, and a construction based on general assumptions. In: Biham, E. (ed.) EUROCRYPT 2003. LNCS, vol. 2656, pp. 614–629. Springer, Heidelberg (2003). https://doi.org/10.1007/3-540-39200-9_38
8. Bettaieb, S., Bidoux, L., Blazy, O., Connan, Y., Gaborit, P.: A gapless code-based hash proof system based on RQC and its applications. Cryptology Eprint https://eprint.iacr.org/2021/026
9. Ezerman, M.F., Lee, H.T., Ling, S., Nguyen, K., Wang, H.: A provably secure group signature scheme from code-based assumptions. In: Iwata, T., Cheon, J.H. (eds.) ASIACRYPT 2015. LNCS, vol. 9452, pp. 260–285. Springer, Heidelberg (2015)
10. Bardet, M., et al.: Algebraic attacks for solving the rank decoding and Minrank problems without Gröbner basis (2020). https://arxiv.org/pdf/2002.08322.pdf
11. Debris-Alazard, T., Tillich, J.-P.: Two attacks on rank metric code-based schemes: RankSign and an identity-based-encryption scheme. In: Peyrin, T., Galbraith, S. (eds.) ASIACRYPT 2018. LNCS, vol. 11272, pp. 62–92. Springer, Cham (2018). https://doi.org/10.1007/978-3-030-03326-2_3
12. Finiasz, M., Sendrier, N.: Security bounds for the design of code-based cryptosystems. In: Matsui, M. (ed.) ASIACRYPT 2009. LNCS, vol. 5912, pp. 88–105. Springer, Heidelberg (2009). https://doi.org/10.1007/978-3-642-10366-7_6
13. Gaborit, P., Schrek, J., Zémor, G.: Full cryptanalysis of the Chen identification protocol. In: Yang, B.-Y. (ed.) PQCrypto 2011. LNCS, vol. 7071, pp. 35–50. Springer, Heidelberg (2011). https://doi.org/10.1007/978-3-642-25405-5_3
14. Langlois, A., Ling, S., Nguyen, K., Wang, H.: Lattice-based group signature scheme with verifier-local revocation. In: Krawczyk, H. (ed.) PKC 2014. LNCS, vol. 8383, pp. 345–361. Springer, Heidelberg (2014). https://doi.org/10.1007/978-3-642-54631-0_20
15. Nguyen, P.Q., Zhang, J., Zhang, Z.: Simpler efficient group signatures from lattices. In: Katz, J. (ed.) PKC 2015. LNCS, vol. 9020, pp. 401–426. Springer, Heidelberg (2015). https://doi.org/10.1007/978-3-662-46447-2_18

16. Nojima, R., Imai, H., Kobara, K., Morozov, K.: Semantic security for the McEliece cryptosystem without random oracles. Des. Codes Cryptogr. **49**(1–3), 289–305 (2008)
17. Pointcheval, D., Vaudenay, S.: On provable security for digital signature algorithms. Technical report LIENS-96-17, Laboratoire d'Informatique de Ecole Normale Superieure (1997)
18. Stern, J.: A new paradigm for public key identification. IEEE Trans. Inf. Theory **42**(6), 1757–1768 (1996)

The Rank-Based Cryptography Library

Nicolas Aragon[1], Slim Bettaieb[2], Loïc Bidoux[2,3(✉)], Yann Connan[1,2],
Jérémie Coulaud[2], Philippe Gaborit[1], and Anaïs Kominiarz[2]

[1] University of Limoges, Limoges, France
[2] Worldline, Seclin, France
loic.bidoux@owndata.org
[3] Technology Innovation Institute, Abu Dhabi, UAE

Abstract. Rank-based cryptography provides cryptosystems that aim
to be secure against both classical and quantum computers. In the past
few years, the interest for code-based cryptography in the rank metric setting has tremendously increased notably since the beginning of the NIST
post-quantum cryptography standardization process. This paper introduces RBC a library dedicated to Rank-Based Cryptography and details
its design and architecture. The performances of RBC are illustrated
against comparable state of the art librairies. RBC greatly outperforms
those libraries as it is 2 to 5 times faster than NTL and 40 to 138 times
faster than $\mathtt{mpF}_q$ on the multiplication and inversion over $\mathbb{F}_{q^m}^n$ which are
the most critical operations when it comes to rank-based cryptography
performances. In addition, the performances of ROLLO and RQC two
rank-based cryptosystems provided by the library are reported for two
platforms: a desktop computer equipped with an Intel Skylake-X CPU
and an ARM Cortex-M4 microcontroller.

Keywords: RBC · Rank metric · Library · Code-based cryptography

Introduction

Post-quantum cryptography aims at proposing schemes that provide security
against adversaries having access to both classical and quantum computers.
Since the seminal work of McEliece in 1978 [McE78], code-based cryptography using the Hamming metric has established itself as a serious alternative
to classical cryptography. It is based on the difficulty of the syndrome decoding (SD) problem which has been proven NP-complete [BMVT78]. Many code-
based cryptosystems have been proposed over the years culminating during
the NIST post-quantum standardization process [Nis16] whose round 3 features three code-based key encapsulation mechanism (KEM) using the Hamming metric [ABC+20, AMAB+20b, AMAB+20a]. Introduced in 1985 [Gab85],
the rank-metric constitutes a promising avenue for code-based cryptography.
The security of rank-based cryptography relies on the rank syndrome decoding
(RSD) problem which is the rank analogue of the syndrome decoding problem. One of the main benefits of the rank metric is that the time complexity of the best known attacks against the RSD grows faster with respect to

© Springer Nature Switzerland AG 2022
A. Wachter-Zeh et al. (Eds.): CBCrypto 2021, LNCS 13150, pp. 22–41, 2022.
https://doi.org/10.1007/978-3-030-98365-9_2

the size of parameters than for the Hamming metric. As a consequence, rank-based cryptosystems feature smaller ciphertext and key sizes than their Hamming counterpart for identical security level. Rank-based schemes have also been considered in the NIST post-quantum standardization process whose round 2 includes two KEM namely ROLLO [AAB+18, AAB+19a, AAB+20a] and RQC [AAB+17b, AAB+19b, AAB+20b].

In this paper, we introduce RBC [AB+] a new C library dedicated to rank-based cryptography which aims to promote and foster community efforts on code-based cryptography in the rank metric setting. Rank-based cryptography relies on binary field arithmetic for which there already exist several libraries in the literature. Nevertheless, none of these libraries is entirely suitable for our purpose as they don't provide all the functionalities required by rank-based cryptography. Indeed, in order to implement rank-based schemes, one needs functions performing arithmetic of $\mathbb{F}_{q^m}$ elements, arithmetic of polynomials and vector spaces over $\mathbb{F}_{q^m}$ as well as specific functions dedicated to the notion of rank weight. In addition, existing libraries are not satisfactory when it comes to performances. Some libraries are really efficient for arithmetic in $\mathbb{F}_{q^m}$ while other really shine on arithmetic in $\mathbb{F}_{q^m}^n$ unfortunately no existing library is clearly superior to another one when the whole spectrum of rank-based cryptography is considered. Besides, some libraries relies on algorithms that are not the most efficient ones for the values of m and n typically used in rank-based cryptography as they target other applications. All these considerations have motivated the design and release of the RBC library.

Paper Organization. In Sect. 1, we introduce the rank metric, Gabidulin and LRPC codes as well as the ROLLO and RQC cryptosystems. Next in Sect. 2, we describe the design and the architecture of our new library. We also detail some of the algorithms provided by the library focusing on the most critical ones with respect to performances. In Sect. 3, we present the performances of our library by comparing it to the $\mathtt{mpF}_q$, NTL and RELIC libraries. We also showcase its performances by reporting the execution timing of ROLLO and RQC on two platforms: a desktop computer equipped with a Skylake-X CPU and an ARM Cortex-M4 microcontroller. To finish, ongoing and future works related to the RBC library are discussed.

1 Preliminaries

1.1 Rank Metric Overview

The rank metric has been introduced by Gabidulin in 1985 [Gab85]. Let q be a power of a prime p, m be an integer, $\mathbb{F}_{q^m}$ a finite field, $\mathcal{B} = \{\beta_1, \ldots, \beta_m\}$ a basis of $\mathbb{F}_{q^m}$ viewed as a m-dimensional vector space over $\mathbb{F}_q$ and $\mathcal{V}$ a n-dimensional vector space over $\mathbb{F}_{q^m}$. One can express the coordinates of $\mathbf{x} \in \mathcal{V}$ in $\mathcal{B}$ thus defining the matrix $\mathbf{M_x} \in \mathcal{M}_{m,n}(\mathbb{F}_q)$ where $\mathbf{M_x} = (x_{i,j})$ such that $x_j = \sum_{i=1}^{m} x_{i,j}\beta_i$ for all $j \in [\![0, n-1]\!]$.

$$\mathbf{M}: \qquad \mathbb{F}_{q^m}^n \qquad \simeq \qquad \mathcal{M}_{m,n}(\mathbb{F}_q)$$

$$\mathbf{x} = (x_0, \ldots, x_{n-1}) \mapsto \mathbf{M_x} = \begin{pmatrix} x_{1,0} & \cdots & x_{1,n-1} \\ x_{2,0} & \cdots & x_{2,n-1} \\ \vdots & & \vdots \\ x_{m,0} & \cdots & x_{m,n-1} \end{pmatrix} \begin{matrix} \beta_1 \\ \beta_2 \\ \vdots \\ \beta_m \end{matrix}$$

Let $P \in \mathbb{F}_q[X]$ be a polynomial of degree n, one can also identify the vector space $\mathcal{V}$ to the commutative ring $\mathbb{F}_{q^m}[X]/\langle P \rangle$ where $\langle P \rangle$ denotes the ideal of $\mathbb{F}_{q^m}[X]$ generated by P.

$$\Psi: \qquad \mathbb{F}_{q^m}^n \qquad \simeq \qquad \mathbb{F}_{q^m}[X]/\langle P \rangle$$

$$\mathbf{x} = (x_0, \ldots, x_{n-1}) \mapsto \Psi(\mathbf{x}) = \sum_{i=0}^{n-1} x_i X^i$$

For $\mathbf{x}, \mathbf{y} \in \mathcal{V}$, the product $\mathbf{z} = \mathbf{x} \cdot \mathbf{y}$ is defined using the polynomial multiplication in $\mathbb{F}_{q^m}[X]/\langle P \rangle$ namely $\mathbf{z}$ is the only vector such that $\Psi(\mathbf{z}) = \Psi(\mathbf{x}) \cdot \Psi(\mathbf{y})$. To finish, we introduce the support and rank weight of $\mathbf{x} \in \mathcal{V}$ which are two core notions in rank-based cryptography.

Definition 1 (Support). *The support of $\mathbf{x} = (x_0, \ldots, x_{n-1}) \in \mathcal{V}$, denoted $\mathsf{Supp}(\mathbf{x})$, is the $\mathbb{F}_q$-subspace of $\mathbb{F}_{q^m}$ generated by the coordinates of $\mathbf{x}$ namely $\mathsf{Supp}(\mathbf{x}) = \langle x_0, \ldots, x_{n-1} \rangle_{\mathbb{F}_q}$.*

Definition 2 (Rank weight). *The rank weight of $\mathbf{x} = (x_0, \ldots, x_{n-1}) \in \mathcal{V}$, denoted $\|\mathbf{x}\|$, is defined as the dimension of $\mathsf{Supp}(\mathbf{x})$ or equivalently as the rank of the matrix $\mathbf{M_x}$.*

1.2   Rank Metric Codes

There are two main families of codes in rank metric. Gabidulin codes [Gab85] are analogue to the Reed-Solomon codes and can be thought as the evaluation of q-polynomials [Ore33] of bounded degree on the coordinates of a vector over $\mathbb{F}_{q^m}$. Gabidulin codes are $\mathbb{F}_{q^m}$-linear codes that can deterministically decode up to $\lfloor \frac{n-k}{2} \rfloor$ errors. Low Rank Parity Check (LRPC) codes [GMRZ13] are $\mathbb{F}_{q^m}$-linear codes whose parity check matrix coefficients belong to a space of small dimension. Unlike Gabidulin codes, LRPC codes are probabilistic and as such they feature a non-zero decoding failure probability.

Definition 3 ($\mathbb{F}_{q^m}$-linear code). *An $\mathbb{F}_{q^m}$-linear code $\mathcal{C}$ of dimension k and length n, denoted $[n, k]_{q^m}$, is a subspace of $\mathbb{F}_{q^m}^n$ of dimension k.*

Definition 4 (Generator matrix). *A matrix $\mathbf{G} \in \mathbb{F}_{q^m}^{m \times n}$ is a generator matrix for the $[n, k]_{q^m}$ code $\mathcal{C}$ if $\mathcal{C} = \{\mathbf{xG} \mid \mathbf{x} \in \mathbb{F}_{q^m}^k\}$.*

Definition 5 (Parity-check matrix). *A matrix $\mathbf{H} \in \mathbb{F}_{q^m}^{(n-k) \times n}$ is a parity-check matrix for the $[n, k]_{q^m}$ code $\mathcal{C}$ if $\mathcal{C} = \{\mathbf{x} \in \mathbb{F}_{q^m}^n \mid \mathbf{Hx}^\top = 0\}$. The vector $\mathbf{Hx}^\top \in \mathbb{F}_{q^m}^{n-k}$ is called the syndrome of $\mathbf{x}$.*

Definition 6 (*q-polynomials*). *The set of q-polynomials over* $\mathbb{F}_{q^m}$ *is the set of polynomials with the following shape:* $\{P(X) = \sum_{i=0}^{r} p_i X^{q^i} \mid p_i \in \mathbb{F}_{q^m}, p_r \neq 0\}$. *The q-degree of a q-polynomial P is defined as* $deg_q(P) = r$.

Definition 7 (**Gabidulin codes**). *Let* $k, n, m \in \mathbb{N}$ *such that* $k \leqslant n \leqslant m$. *Let* $\mathbf{g} = (g_0, \ldots, g_{n-1})$ *be a* $\mathbb{F}_q$-*linearly independent family of vectors of* $\mathbb{F}_{q^m}$. *The Gabidulin code* $\mathcal{G}_g(n, k, m)$ *is the code defined as* $\{P(\mathbf{g}) \mid deg_q(P) < k\}$ *where* $P(\mathbf{g}) := (P(g_1), \ldots, P(g_n))$.

Definition 8 (**LRPC codes**). *Let* $\mathbf{H} = (h_{ij})_{i \in [\![1, n-k]\!], \, j \in [\![1, n]\!]} \in \mathbb{F}_{q^m}^{(n-k) \times n}$ *be a full-rank matrix such that its coefficients generate an* $\mathbb{F}_q$-*subspace F of small dimension d, i.e.* $F = \langle h_{ij} \rangle_{\mathbb{F}_q}$ *and* $d = dim(F)$. *Let C be the code with parity-check matrix* $\mathbf{H}$, *C is called an* $[n, k]_{q^m}$ *LRPC code.*

1.3 The ROLLO and RQC Schemes

ROLLO. ROLLO is the merge of the three cryptosystems LAKE [ABD+17a], LOCKER [ABD+17b] and Rank-Ouroboros [AAB+17a] which all share the same decryption algorithm for LRPC codes. Following [AAB+20a], we only consider ROLLO-I (formerly LAKE) and ROLLO-II (formerly LOCKER) in the remaining of this paper. ROLLO-I is an IND-CPA KEM whereas ROLLO-II is an IND-CCA2 public key encryption (PKE) scheme. They are respectively described in Fig. 1 and Fig. 2 from Appendix A; we defer the interested reader to [AAB+20a] for additional details.

RQC. RQC is an IND-CCA2 KEM build from an IND-CPA PKE construction on top of which the HHK transform [HHK17] is performed. Unlike many other code-based cryptosystems, the security of RQC does not rely on any code indistinguishability assumption following the approach introduced by Alekhnovich [Ale03]. We only describe the PKE version of RQC for simplicity (see Appendix A, Fig. 3) and defer the reader to [AAB+20b] for additional details.

2 The RBC Library

In this section, we describe the design and the architecture of our new library (Sects. 2.1 and 2.2). We also detail some algorithms provided by the library focusing on the most critical ones with respect to performances (Sect. 2.3).

2.1 RBC Library Overview

RBC [AB+] is a C library dedicated to rank-based cryptography that focuses on performances without sacrificing usability. It is released under the LGPL license and can be retrieved at https://rbc-lib.org. It currently features:

- A *core layer* providing arithmetic for elements, vectors and polynomials over $\mathbb{F}_{2^m}$ with some utility functions tailored to rank-based cryptography;

- A *code layer* providing implementations for the main codes used in rank-based cryptography namely Gabidulin codes and LRPC codes;
- A *scheme layer* providing implementations for ROLLO and RQC, two rank-based cryptosystems submitted to the NIST PQC standardization process.

Dual API. The RBC library API can be thought as a dual API targeting two different audiences. We refer as *end users* people who are mainly concerned with using the schemes provided by the library (for instance to include ROLLO in a software, benchmark rank-metric based cryptosystems...) and we refer as *advanced users* people who want to use rank-based cryptography functionalities that are not limited to the schemes provided by the library (for instance to implement a new rank-metric based cryptosystem, contribute to the library...). End users should consider that the RBC library API is limited to the scheme layer functions while advanced users should use the whole API namely functions from the core, code and scheme layers.

Design Choice Regarding Finite Fields. The RBC library currently only supports finite fields of the form $\mathbb{F}_{q^m}$ with $q = 2$ which are the most commonly used finite fields in rank-metric cryptography. Regarding implementation of finite field arithmetic, one can either provide generic algorithms suited for any value of m or provide specific algorithms tailored for each value of m. While the first approach is superior with respect to simplicity and usability, the RBC library uses the second approach which is better when it comes to performances. This has no impact on usability for end users but adds some complexity for advanced users which is partly mitigated thanks to our preprocessing and build system.

Preprocessing and Build System. RBC library preprocessing and build system is a set of python scripts facilitating development for advanced users and allowing build customization for all users. It features a templating system for the core layer that generates optimized code for each finite field while avoiding code redundancy. In addition, it provides automatic source code specialization for the code and scheme layers allowing users to write generic code that will be automatically instantiated with finite fields specified in a configuration file. Doing so, one can write generic code while keeping the possibility to use several instantiations of its code at once. For instance, writing only one ROLLO-I implementation and creating a program that call both ROLLO-I-128 and ROLLO-I-192 instantiated respectively with $\mathbb{F}_{2^{67}}^{83}$ and $\mathbb{F}_{2^{79}}^{97}$ while avoiding any code redundancy. In addition, the RBC preprocessing and build system allows users to customize the build of the library by specifying several options in a configuration file. Users may choose the targeted architecture amongst x86, x64 and x64 along with CLMUL and AVX2 support. The preprocessing and build system will generate code accordingly by choosing the best available algorithms for the specified architecture. Users may also choose which cryptosystems from the scheme layer they want to include in their build thus offering the possibility to minimize the size of the generated library files.

Tests, Documentation and Examples. In order to ease the use of the RBC library, a documentation is available. In addition, working examples and benchmark tools are provided for the cryptosystems included in the library. Unit-tests are available for the core layer functions and KAT tests are provided for the code and scheme layers.

Third-Party Implementations. The RBC library relies on several crypto-graphic primitives that are outside the scope of rank-based cryptography such as a pseudorandom number generator, a seedexpander, SHA2, FIPS202 and AES. Implementations for theses primitives are retrieved from the BearSSL [Por16], OpenSSL [Ope], PQClean [PQC], SUPERCOP [Sup] projects and [Nis16,Gue10]. In addition, the Minunit framework [Min] and the $\text{mp}\mathbb{F}_q$ library [GT07,GT08] are used to provide unit-tests against the library.

2.2   RBC Library Architecture

The RBC library introduces several structures and types corresponding to math-ematical objects manipulated in rank-based cryptography. They are easily iden-tified thanks to their common rbc prefix.

The following structures constitute the core layer of the RBC library:

- rbc_elt implementing an element of $\mathbb{F}_{q^m}$;
- rbc_vec implementing a vector over $\mathbb{F}_{q^m}$;
- rbc_vspace implementing a vector space over $\mathbb{F}_{q^m}$;
- rbc_poly implementing a polynomial over $\mathbb{F}_{q^m}$;
- rbc_qre implementing an element of the quotient ring $\mathbb{F}_{q^m}[X]/\langle P \rangle$ where $\langle P \rangle$ denotes the ideal of $\mathbb{F}_{q^m}[X]$ generated by P.

These types have various dependencies one to each other. For instance, rbc_vec are constructed from rbc_elt while rbc_vspace and rbc_poly are based on rbc_vec. In addition, the rbc_qre type is built from the rbc_poly one. For each of the aforementioned types, the library provides arithmetic operations, generation of random elements, serialization as well as utility functions tailored to rank-based cryptography.

Additional types and functions are defined within RBC code layer:

- rbc_qpoly implementing a q-polynomial over $\mathbb{F}_{q^m}$;
- rbc_gabidulin implementing a Gabidulin code;
- rbc_lrpc_RSR() providing LRPC decoding.

The rbc_gabidulin type relies on rbc_qpoly in order to provide encoding and decoding algorithms for Gabidulin codes. As LRPC encoding is generally performed using rbc_qre arithmetic, we provide LRPC decoding using only the rbc_lrpc_RSR() function.

The scheme layer follows a different convention where the rbc prefix is replaced by schemeName_securityLevel for convenience. For instance, ROLLO -I-128 can be instantiated using the following functions: rolloI_128_kem_keygen(), rolloI_128_kem_encaps() and rolloI_128_kem_decaps().

2.3   RBC Library Algorithms

In this section, we detail some of the algorithms implemented in the RBC library focusing on the most critical ones with respect to performances. As arithmetic over $\mathbb{F}_{q^m}^n$ is of paramount importance in rank metric, we have selected algorithms that are well suited for the values of m and n typically used in rank-based cryptography. Hereafter, we denote by the *RBC supported instructions sets* the CLMUL and AVX2 instruction sets. For some operations, we provide two implementations depending on whether the RBC supported instruction sets can be used or not. These instructions are leveraged using Intel intrinsics therefore we refer to them with the name of the corresponding intrinsics instruction.

Algorithms Related to rbc_elt
The RBC library uses polynomial representation for the rbc_elt therefore elements $e \in \mathbb{F}_{2^m}$ are represented as vectors $(e_0, \ldots, e_{m-1})$ of size m over $\mathbb{F}_2$. Operations in $\mathbb{F}_{2^m}$ are performed using polynomial arithmetic modulo Π where Π is the sparse irreducible polynomial used to define $\mathbb{F}_{2^m}$ as $\mathbb{F}_2[X]/\langle\Pi\rangle$. As such, many operations on rbc_elt generate unreduced elements (represented by the rbc_elt_ur type) that can be transformed to rbc_elt by performing reduction modulo Π.

Multiplication. The rbc_elt_mul() function encompasses a polynomial multiplication followed by a modular reduction. Two polynomial multiplication algorithms are provided depending on whether the RBC supported instruction sets can be used or not. If the aforementioned instruction sets are supported, a textbook polynomial multiplication accelerated by the _mm_clmulepi64() intrinsics instruction is performed. Otherwise, the multiplication is implemented using the left-to-right comb method with preprocessing; see Algorithm 2.36 of [HMV06] for additional details.

Inversion. The rbc_elt_inv() function is implemented using a version of the Euclidean algorithm tailored for binary fields.

Squaring. The rbc_elt_sqr() function inserts several zeros within the representation of an element $e = (e_0, e_1, \ldots, e_{m-1})$ in order to obtain an unreduced element $e' = (e_0, 0, e_1, 0, \ldots, e_{m-1})$ which correspond to squaring in $\mathbb{F}_{2^m}$ after modular reduction of e'. If the RBC supported instruction sets are available, this is done using the interleaving intrinsics instructions _mm_unpacklo_epi8() and _mm_unpackhi_epi8() along with preprocessing; see Algorithm 1 of [ALH10]. A similar but less efficient algorithm is used if the aforementioned instruction sets are not supported.

Modular Reduction. The rbc_elt_reduce() function uses an algorithm that exploits the sparse structure of the polynomial Π by performing reduction over $\mathbb{F}_{2^m}$ one word at a time. This algorithm is tailored to each considered finite field as Π differs for each value of m; see Figure 2.9 and Algorithm 2.41 of [HMV06] for an example over $\mathbb{F}_{2^{163}}$.

Algorithms Related to rbc_vec

The rbc_vec is an utility type mainly used to construct the rbc_poly and rbc_vspace types nevertheless it provides some core functionalities for rank-based cryptography such as random vectors generation and rank weight computation. It is implemented as a pointer of rbc_elt whose size is fixed at initialization without any resize function provided.

Random Vectors Generation. Three ways of generating random vectors over $\mathbb{F}_{q^m}$ are provided in the RBC library:

1. The rbc_vec_set_random() function generates a vector purely at random by sampling each of its coordinate randomly in $\mathbb{F}_{q^m}$;
2. The rbc_vec_set_random_full_rank() function generates a full rank vector. To do so, each coordinates of the vector is sampled randomly in $\mathbb{F}_{q^m}$ then the rank weight of the vector is computed. This process is repeated until the vector is of full rank;
3. The rbc_vec_set_random_from_support() function generates a vector randomly with each coordinate sampled from a support F of dimension d. First, the generating family of F is copied at random positions of the vector then the remaining coordinates are filled with random linear combinations of the generating family of F.

Rank Weight. The rbc_vec_get_rank() function determines the rank weight of a vector $\mathbf{x} \in \mathbb{F}_{2^m}^n$ by computing the rank of its associated matrix $\mathbf{M_x}$ using the Gauss algorithm.

Algorithms Related to rbc_poly

The rbc_poly type is implemented as a structure containing a rbc_vec element used to store the coefficients of the polynomial, the current **degree** of the polynomial and a **max_degree** value that keeps track of the size of the underlying rbc_vec element.

Multiplication. The rbc_poly_mul() function implements a recursive Karatsuba algorithm. Each level of recursion splits each of the polynomials in half and an hardcoded multiplication is used when the degrees of both polynomials is at most one. Our implementation is inspired from the NTL library [S+01].

Inversion. The rbc_poly_inv() function implements polynomial inversion using the extended Euclidean algorithm.

Algorithms Related to rbc_vspace

Vector spaces are represented using generating families therefore the rbc_vspace type is simply a rbc_vec and the corresponding subspace of $\mathbb{F}_{q^m}$ is the vector space generated by the elements stored within the rbc_vec.

Direct Sum. The rbc_vspace_directsum() function computes the direct sum of two vector spaces A and B by concatenating their generating families.

Product. Given a vector spaces A and B of generating families $(A_0, \ldots, A_{d-1})$ and $(B_0, \ldots, B_{r-1})$, the `rbc_vspace_product()` function calculates their product C of generating family $(C_{0,0}, \ldots, C_{r-1,d-1})$ by computing the following elements: $C_{i,j} = A_i \times B_j$ for $i \in [0, d-1]$ and $j \in [0, r-1]$.

Intersection. The `rbc_vspace_intersection()` function computes the intersection of two vector spaces A and B by using the Zassenhaus algorithm.

Canonical Basis. Some cryptosystems use vector spaces as inputs to hash functions therefore one needs to be able to represent vector spaces in a non ambiguous way. Given a vector space V represented by a `rbc_vec` **v**, one can compute the row echelon form of the matrix $\mathbf{M_v}$ associated to **v** by calling the `rbc_vec_echelonize()` function thus obtaining a canonical basis of V.

Algorithms Related to Gabidulin and LRPC Codes

Gabidulin. The `rbc_gabidulin_encode()` function performs Gabidulin codes encoding using a classical vector/matrix multiplication. Gabidulin codes decoding is realized by the `rbc_gabidulin_decode()` function using the algorithm proposed by Loidreau in [Loi05] and later improved in [ALR18]. This algorithm uses the q-polynomial reconstruction method and as such relies extensively on the arithmetic of the ring of q-polynomials over $\mathbb{F}_{2^m}$ which is provided by the `rbc_qpoly` structure. More precisely, the RBC library implement the variant described in [BBGM19] along with the "Polynomials with lower degree" optimization from section 4.4.2 of [ALR18].

LRPC. No specific structure for LRPC codes have been provided as LRPC encoding is generally performed throught `rbc_qre` arithmetic. LRPC decoding is performed by the `rbc_lrpc_RSR()` function that implements the Rank Support Recover algorithm; see Algorithm 1 of [AAB+18]. This algorithm is similar to the standard LRPC codes decoding algorithm described in [GMRZ13] except that it stops after recovering the support E of the error vector **e**.

3   RBC Library Performances

In this section, we discuss the performances of the RBC library by comparing it to the $\mathtt{mp}\mathbb{F}_q$, NTL and RELIC libraries (Sect. 3.1). Next, we report the performances of RQC and ROLLO as implemented in the library for two platforms: a desktop computer equipped with a Skylake-X CPU (Sect. 3.2) and a Cortex-M4 microcontroller (Sect. 3.3).

3.1   Comparison with the NTL, $\mathtt{mp}\mathbb{F}_q$ and RELIC libraries

The benchmarks have been performed on a machine that has 16 GB of memory and an Intel® Core™ i7-7820X (Skylake-X) CPU @ 3.6 GHz for which the Hyper-Threading, Turbo Boost and SpeedStep features were disabled. The following libraries have been used: NTL [S+01] (version 11.4.3) along with GF2X (version 1.3.0) and GMP (version 6.2.0), $\mathtt{mp}\mathbb{F}_q$ [GT07, GT08] (version 1.1) and

RELIC [AGM+] (version 0.5.0). The benchmarks have been compiled with GCC (version 10.1.0) using the -O3 -flto -mavx2 -mpclmul -msse4.2 -maes flags. The results have been obtained by computing the average running time from 1000 random instances. In order to minimize biases from background tasks running on the benchmark platform, each instance have been repeated 100 times and averaged. Our benchmark is focused on the finite fields corresponding to the different parameters sets of ROLLO and RQC. The RELIC library provides several implementations for each arithmetic operation; we have tested all implementations while reporting only the most efficient one.

Multiplication and inversion over $\mathbb{F}_{q^m}^n$ (namely operations over $\mathbb{F}_{q^m}[X]/P$ for some polynomial P) are the most critical operations when it comes to rank-based cryptography performances. One can see from Table 1 below (as well as Appendix B, Tables 2, 3, 4, 5, 6, 7, 8, 9 and 10) that RBC greatly outperforms other libraries on these operations as it is 2 to 5 times faster than NTL and 40 to 138 times faster than mp$\mathbb{F}_q$. Overall, the RBC library is more efficient than NTL and RELIC on all the considered operations nevertheless mp$\mathbb{F}_q$ sometimes outperforms RBC on arithmetic operations over $\mathbb{F}_{q^m}$. Indeed, one can see that inversion over $\mathbb{F}_{q^m}$ is about 20 % faster in mp$\mathbb{F}_q$ than in RBC. Multiplication and squaring over $\mathbb{F}_{q^m}$ feature similar performances in RBC and mp$\mathbb{F}_q$ although mp$\mathbb{F}_q$ seems more efficient than RBC when the polynomial used for reduction is a pentanomial. However, whenever $m \geq 128$, RBC outperforms mp$\mathbb{F}_q$ for both multiplication and squaring. Exploring the algorithmic differences between RBC and mp$\mathbb{F}_q$ in the case where the polynomial used for reduction is a pentanomial constitutes an interesting point to explore for future work that may lead to performance improvements.

Table 1. Performances in CPU cycles for $\mathbb{F}_{2^{127}}^{113}$ (RQC-128 parameters)

Operation	RBC	mp$\mathbb{F}_q$	NTL	RELIC
Multiplication over $\mathbb{F}_{2^{127}}$	32	32	223	1 118
Inversion over $\mathbb{F}_{2^{127}}$	5 320	3 924	7 296	7 822
Squaring over $\mathbb{F}_{2^{127}}$	32	90	161	166
Multiplication over $\mathbb{F}_{2^{127}}^{113}$	88 221	8 868 521	453 234	-
Inversion over $\mathbb{F}_{2^{127}}^{113}$	1 548 059	-	5 604 693	-

3.2 Performances of ROLLO and RQC on Intel Skylake-X

The benchmarks have been performed on a machine that has 16 GB of memory and an Intel® Core™ i7-7820X (Skylake-X) CPU @ 3.6 GHz for which the Hyper-Threading, Turbo Boost and SpeedStep features were disabled. The schemes have been compiled with GCC (version 10.1.0) using the -O3 -flto

-mavx2 -mpclmul -msse4.2 -maes -std=c99 flags. The OpenSSL library (version 1.1.1.g) have been used as a provider for SHA2. The results have been obtained by computing the average running time from 1000 random instances. In order to minimize biases from background tasks running on the benchmark platform, each instance have been repeated 100 times and averaged.

One can see from Appendix B, Tables 11, 12 and 13 that ROLLO and RQC are both efficient on the x64 architecture. Indeed, one can compute the Keygen, Encaps and Decaps operations of ROLLO-I-128 and ROLLO-I-256 in respectively less than 0.5 ms and 1 ms on our benchmark machine. ROLLO-II is slightly less efficient as an inversion over $\mathbb{F}_{q^m}^n$ have to be performed during the Keygen. Nonetheless all the operations of ROLLO-II-128 and ROLLO-II-256 can be computed in respectively less than 1.5 ms and 2 ms on our benchmark machine. RQC-128 is also fairly efficient as the Keygen, Encaps and Decaps can be computed in less than 1 ms on the considered machine. However, Gabidulin decoding become costly for bigger parameters therefore one need up to 3.5 ms to compute the Keygen, Encaps and Decaps of RQC-256 on our benchmark machine.

3.3   Performances of ROLLO and RQC on ARM Cortex-M4

In this section, we present the performances of ROLLO and RQC as implemented within the RBC library on microcontroller. Several implementations have been reported in the literature. The first one provide an implementation of ROLLO-I leveraging the ARM SecurCore SC300 crypto co-processor [LMB+19] while the second one studies the Encaps operation of both ROLLO and RQC on the ARM Cortex-M0 microcontroller [ABC+19]. Hereafter, we focus on the ARM Cortex-M4 microcontroller as suggested by the NIST and therefore compare our results to those of the pqm4 project [KRSS20] that aims to provide a post-quantum cryptography library for the Cortex-M4.

The benchmarks have been performed on a STM32F4 discovery board featuring a 32-bit ARM-Cortex-M4 processor, 1 MByte flash memory and 196 KByte RAM. Our tests use the pqm4 benchmark scripts and as such follow the methodology described in [KRSS20]. In particular, all cycle counts are obtained at 24 MHz. For each scheme, 100 executions have been performed using arm-none-eabi-gcc in version 10.1.0. The mean running times for ROLLO-I, ROLLO-II and RQC are presented in Appendix C, Tables 14, 15 and 16. No value is reported for RQC-256 as the current implementation exceeds the available memory of the targeted platform. We defer to future work the design of a memory optimized implementation of RQC-256. In order to contextualize these results, the Table 17 depicts the performances of some post-quantum KEM included in pqm4 focusing on C implementations targeting 128 bits of security (*i.e.* comparable to ROLLO-I-128, ROLLO-II-128 and RQC-128). Out of fairness for projects that have released implementations with Cortex-M4 specific optimizations (which we did not do), we have also reported their performances in Appendix C, Table 18.

Implementations from the pqm4 project are based on the implementations targeting the 64-bit architecture submitted to the NIST PQC standardization process. Hereafter, we report improvements over this work using our new implementations targeting 32-bit architectures. The observed running timings for ROLLO and RQC are up to twice as fast as the ones currently reported in pqm4.

Ongoing and Future Work

The first version of the RBC library constitutes a solid basis to support people implementing rank-based cryptography. Nonetheless, the RBC library is still in its infancy and will be improved over time. In the short term, our priority is to provide a better treatment of constant-time within the library. While some functionalities have received some attention with respect to constant-time (mostly the rbc_elt operations), others functions from the library may be implemented in a non constant-time way. Our future releases will include improvements with respect to constant-time within the library (by considering results from [AMADG21, ABC+] for example as well as improving the rbc_vec, rbc_poly and rbc_vspace operations).

Some avenues worth exploring for future work include (somewhat sorted by priority): (i) integrating additional rank-based cryptosystems such as Durandal [ABG+19], (ii) integrating additional finite fields to RBC as the library currently only provides the ones used by ROLLO and RQC, (iii) exploring the algorithmic improvements mentioned in Sect. 3.1 as well as (iv) exploring potential algorithmic improvements using other representations for $\mathbb{F}_{q^m}$ elements such as normal bases. The RBC library aims to promote community efforts on rank-based cryptography and as such contributions are welcomed. People interested to contribute are invited to contact the library authors.

Appendix A ROLLO and RQC

This appendix describes the ROLLO and RQC schemes. Let $\mathcal{S}_w^n(\mathbb{F}_{q^m})$, $\mathcal{S}_{1,w}^n(\mathbb{F}_{q^m})$ and $\mathcal{S}_{(w_1,w_2)}^{3n}(\mathbb{F}_{q^m})$ be defined as:

$$\mathcal{S}_w^n(\mathbb{F}_{q^m}) = \{\mathbf{X} \in \mathbb{F}_{q^m}^n : \|\mathbf{X}\| = w\}$$

$$\mathcal{S}_{1,w}^n(\mathbb{F}_{q^m}) = \{\mathbf{X} \in \mathbb{F}_{q^m}^n : \|\mathbf{X}\| = w, 1 \in \mathsf{Supp}(\mathbf{X})\}$$

$$\mathcal{S}_{(w_1,w_2)}^{3n}(\mathbb{F}_{q^m}) = \{\mathbf{X} = (\mathbf{X}_1, \mathbf{X}_2, \mathbf{X}_3) \in \mathbb{F}_{q^m}^{3n} : \|(\mathbf{X}_1, \mathbf{X}_3)\| = w_1, \|\mathbf{X}_2\| = w_1 + w_2,$$
$$\mathsf{Supp}(\mathbf{X}_1, \mathbf{X}_3) \subset \mathsf{Supp}(\mathbf{X}_2)\}$$

⋄ Setup(1^λ): generates and outputs the global parameters param = (n, m, d, r, P) where $P \in \mathbb{F}_q[X]$ is an irreducible polynomial of degree n.

⋄ KeyGen(param): Picks $(\mathbf{x}, \mathbf{y}) \xleftarrow{\$} \mathcal{S}_d^{2n}(\mathbb{F}_{q^m})$. Sets $\mathbf{h} = \mathbf{x}^{-1}\mathbf{y} \bmod P$ and returns pk = $\mathbf{h}$ and sk = $(\mathbf{x}, \mathbf{y})$.

⋄ Encaps(pk): Picks $(\mathbf{e}_1, \mathbf{e}_2) \xleftarrow{\$} \mathcal{S}_r^{2n}(\mathbb{F}_{q^m})$, sets $E = \mathsf{Supp}(\mathbf{e}_1, \mathbf{e}_2)$, $\mathbf{c} = \mathbf{e}_1 + \mathbf{e}_2 \cdot \mathbf{h} \bmod P$. Computes $K = \mathsf{Hash}(E)$ and returns $\mathbf{c}$.

⋄ Decaps(sk, $\mathbf{c}$): Sets $\mathbf{s} = \mathbf{x} \cdot \mathbf{c} \bmod P$, $F = \mathsf{Supp}(\mathbf{x}, \mathbf{y})$ and $E = \mathsf{RSR}(F, \mathbf{s}, r)$. Computes $K = \mathsf{Hash}(E)$.

Fig. 1. Description of ROLLO-I [AAB+20a]

⋄ Setup(1^λ): generates and outputs the global parameters param = (n, m, d, r, P) where $P \in \mathbb{F}_q[X]$ is an irreducible polynomial of degree n.

⋄ KeyGen(param): Picks $(\mathbf{x}, \mathbf{y}) \xleftarrow{\$} \mathcal{S}_d^{2n}(\mathbb{F}_{q^m})$. Sets $\mathbf{h} = \mathbf{x}^{-1}\mathbf{y} \bmod P$ and returns pk = $\mathbf{h}$ and sk = $(\mathbf{x}, \mathbf{y})$.

⋄ Encrypt(μ, pk): Picks $(\mathbf{e}_1, \mathbf{e}_2) \xleftarrow{\$} \mathcal{S}_r^{2n}(\mathbb{F}_{q^m})$, sets $E = \mathsf{Supp}(\mathbf{e}_1, \mathbf{e}_2)$, $\mathbf{c} = \mathbf{e}_1 + \mathbf{e}_2 \cdot \mathbf{h} \bmod P$. Computes $c' = \mu \oplus \mathsf{Hash}(E)$ and returns the ciphertext $C = (\mathbf{c}, c')$.

⋄ Decrypt(C, sk): Sets $\mathbf{s} = \mathbf{x} \cdot \mathbf{c} \bmod P$, $F = \mathsf{Supp}(\mathbf{x}, \mathbf{y})$ and $E = \mathsf{RSR}(F, \mathbf{s}, r)$. Returns $\mu = c' \oplus \mathsf{Hash}(E)$.

Fig. 2. Description of ROLLO-II [AAB+20a]

⋄ Setup(1^λ): generates and outputs the global parameters param = $(n, k, \delta, w, w_1, w_2, P)$ where $P \in \mathbb{F}_q[X]$ is an irreducible polynomial of degree n.

⋄ KeyGen(param): Samples $\mathbf{h} \xleftarrow{\$} \mathbb{F}_{q^m}^n$, $\mathbf{g} \xleftarrow{\$} \mathcal{S}_n^n(\mathbb{F}_{q^m})$, $(\mathbf{x}, \mathbf{y}) \xleftarrow{\$} \mathcal{S}_{1,w}^{2n}(\mathbb{F}_{q^m})$, computes the generator matrix $\mathbf{G} \in \mathbb{F}_{q^m}^{k \times n}$ of $\mathcal{G}_\mathbf{g}(n, k, m)$, sets pk = $(\mathbf{g}, \mathbf{h}, \mathbf{s} = \mathbf{x} + \mathbf{h} \cdot \mathbf{y} \bmod P)$ and sk = $(\mathbf{x}, \mathbf{y})$, returns (pk, sk).

⋄ Encrypt(pk, μ, θ): uses randomness θ to generate $(\mathbf{r}_1, \mathbf{e}, \mathbf{r}_2) \xleftarrow{\$} \mathcal{S}_{(w_1, w_2)}^{3n}(\mathbb{F}_{q^m})$, sets $\mathbf{u} = \mathbf{r}_1 + \mathbf{h} \cdot \mathbf{r}_2 \bmod P$ and $\mathbf{v} = \mathbf{mG} + \mathbf{s} \cdot \mathbf{r}_2 + \mathbf{e} \bmod P$, returns $\mathbf{c} = (\mathbf{u}, \mathbf{v})$.

⋄ Decrypt(sk, $\mathbf{c}$): returns $\mathcal{G}_\mathbf{g}.\mathsf{Decode}(\mathbf{v} - \mathbf{u} \cdot \mathbf{y} \bmod P)$.

Fig. 3. Description of the PKE version of RQC [AAB+20b]

Appendix B RBC Library Performances

Table 2. Performances in CPU cycles for $\mathbb{F}_{2^{67}}^{83}$ (ROLLO-I-128 parameters)

Operation	RBC	mp$\mathbb{F}_q$	NTL	RELIC
Multiplication over $\mathbb{F}_{2^{67}}$	60	32	448	1 175
Inversion over $\mathbb{F}_{2^{67}}$	2 670	2 327	4 099	4 347
Squaring over $\mathbb{F}_{2^{67}}$	60	32	406	208
Multiplication over $\mathbb{F}_{2^{67}}^{83}$	73 821	3 220 639	316 721	-
Inversion over $\mathbb{F}_{2^{67}}^{83}$	771 595	-	6 554 298	-

Table 3. Performances in CPU cycles for $\mathbb{F}_{2^{79}}^{97}$ (ROLLO-I-192 parameters)

Operation	RBC	mp$\mathbb{F}_q$	NTL	RELIC
Multiplication over $\mathbb{F}_{2^{79}}$	31	32	218	1 161
Inversion over $\mathbb{F}_{2^{79}}$	3 147	2 529	5 010	5 019
Squaring over $\mathbb{F}_{2^{79}}$	31	32	159	167
Multiplication over $\mathbb{F}_{2^{79}}^{97}$	79 367	4 801 501	370 442	-
Inversion over $\mathbb{F}_{2^{79}}^{97}$	989 418	-	4 889 575	-

Table 4. Performances in CPU cycles for $\mathbb{F}_{2^{97}}^{113}$ (ROLLO-I-256 parameters)

Operation	RBC	mp$\mathbb{F}_q$	NTL	RELIC
Multiplication over $\mathbb{F}_{2^{97}}$	32	32	225	1 094
Inversion over $\mathbb{F}_{2^{97}}$	4 036	3 081	5 891	6 004
Squaring over $\mathbb{F}_{2^{97}}$	32	32	187	166
Multiplication over $\mathbb{F}_{2^{97}}^{113}$	87 471	7 607 949	353 243	-
Inversion over $\mathbb{F}_{2^{97}}^{113}$	1 403 285	-	6 232 926	-

Table 5. Performances in CPU for $\mathbb{F}_{2^{83}}^{189}$ (ROLLO-II-128 parameters)

Operation	RBC	mp$\mathbb{F}_q$	NTL	RELIC
Multiplication over $\mathbb{F}_{2^{83}}$	57	32	238	1 115
Inversion over $\mathbb{F}_{2^{83}}$	3 384	2 650	5 072	5 303
Squaring over $\mathbb{F}_{2^{83}}$	32	32	192	208
Multiplication over $\mathbb{F}_{2^{83}}^{189}$	235 746	20 555 390	621 525	-
Inversion over $\mathbb{F}_{2^{83}}^{189}$	3 287 743	-	12 547 730	-

Table 6. Performances in CPU cycles for $\mathbb{F}_{2^{97}}^{193}$ (ROLLO-II-192 parameters)

Operation	RBC	mp$\mathbb{F}_q$	NTL	RELIC
Multiplication over $\mathbb{F}_{2^{97}}$	32	32	225	1 094
Inversion over $\mathbb{F}_{2^{97}}$	4 036	3 081	5 891	6 004
Squaring over $\mathbb{F}_{2^{97}}$	32	32	187	166
Multiplication over $\mathbb{F}_{2^{97}}^{193}$	238 539	22 797 360	642 396	-
Inversion over $\mathbb{F}_{2^{97}}^{193}$	3 458 307	-	15 756 672	-

Table 7. Performances in CPU cycles for $\mathbb{F}_{2^{97}}^{211}$ (ROLLO-II-256 parameters)

Operation	RBC	mp$\mathbb{F}_q$	NTL	RELIC
Multiplication over $\mathbb{F}_{2^{97}}$	32	32	225	1 094
Inversion over $\mathbb{F}_{2^{97}}$	4 036	3 081	5 891	6 004
Squaring over $\mathbb{F}_{2^{97}}$	32	32	187	166
Multiplication over $\mathbb{F}_{2^{97}}^{211}$	256 874	27 312 415	764 347	-
Inversion over $\mathbb{F}_{2^{97}}^{211}$	4 042 388	-	14 191 588	-

Table 8. Performances in CPU cycles for $\mathbb{F}_{2^{127}}^{113}$ (RQC-128 parameters)

Operation	RBC	mp$\mathbb{F}_q$	NTL	RELIC
Multiplication over $\mathbb{F}_{2^{127}}$	32	32	223	1 118
Inversion over $\mathbb{F}_{2^{127}}$	5 320	3 924	7 296	7 822
Squaring over $\mathbb{F}_{2^{127}}$	32	90	161	166
Multiplication over $\mathbb{F}_{2^{127}}^{113}$	88 221	8 868 521	453 234	-
Inversion over $\mathbb{F}_{2^{127}}^{113}$	1 548 059	-	5 604 693	-

Table 9. Performances in CPU cycles for $\mathbb{F}_{2^{151}}^{149}$ (RQC-192 parameters)

Operation	RBC	mp$\mathbb{F}_q$	NTL	RELIC
Multiplication over $\mathbb{F}_{2^{151}}$	63	215	231	1 351
Inversion over $\mathbb{F}_{2^{151}}$	7 581	6 488	9 878	10 384
Squaring over $\mathbb{F}_{2^{151}}$	65	100	214	185
Multiplication over $\mathbb{F}_{2^{151}}^{149}$	235 771	28 515 864	871 936	-
Inversion over $\mathbb{F}_{2^{151}}^{149}$	3 552 081	-	10 433 894	-

Table 10. Performances in CPU cycles for $\mathbb{F}_{2^{181}}^{179}$ (RQC-256 parameters)

Operation	RBC	mp$\mathbb{F}_q$	NTL	RELIC
Multiplication over $\mathbb{F}_{2^{181}}$	74	435	285	1 408
Inversion over $\mathbb{F}_{2^{181}}$	9 284	7 961	11 743	12 311
Squaring over $\mathbb{F}_{2^{181}}$	76	114	230	237
Multiplication over $\mathbb{F}_{2^{181}}^{179}$	382 830	52 895 485	1 332 610	-
Inversion over $\mathbb{F}_{2^{181}}^{179}$	5 734 491	-	16 249 680	-

Appendix C ROLLO and RQC Performances

Table 11. Performances of ROLLO-I on intel Skylake-X in CPU cycles

Scheme	Keygen	Encaps	Decaps
ROLLO-I-128	869 509	112 651	736 912
ROLLO-I-192	1 075 191	124 980	834 851
ROLLO-I-256	1 514 003	150 117	1 280 401

Table 12. Performances of ROLLO-II on Intel Skylake-X in CPU cycles

Scheme	Keygen	Encaps	Decaps
ROLLO-II-128	3 619 812	332 877	1 144 540
ROLLO-II-192	3 766 107	338 967	1 256 774
ROLLO-II-256	4 394 490	354 564	1 621 820

Table 13. Performances of RQC on Intel Skylake-X in CPU cycles

Scheme	Keygen	Encaps	Decaps
RQC-128	366 445	530 762	2 581 487
RQC-192	798 057	1 200 596	5 739 349
RQC-256	1 165 492	1 713 963	9 466 386

Table 14. Performances of ROLLO-I on ARM Cortex-M4 in cycles

Scheme	Keygen	Encaps	Decaps
ROLLO-I-128	16 927 603	1 926 332	7 009 943
ROLLO-I-192	22 466 486	2 271 969	7 839 572
ROLLO-I-256	45 424 004	3 769 338	15 039 516

Table 15. Performances of ROLLO-II on ARM Cortex-M4 in cycles

Scheme	Keygen	Encaps	Decaps
ROLLO-II-128	85 063 257	6 844 408	17 321 266
ROLLO-II-192	128 155 854	9 687 469	24 668 141
ROLLO-II-256	152 145 827	10 867 964	29 573 929

Table 16. Performances of RQC on ARM Cortex-M4 in cycles

Scheme	Keygen	Encaps	Decaps
RQC-128	5756 747	11 340 541	71 551 978
RQC-192	12 324 464	24 632 358	150 108 887

Table 17. Performances of several KEM on ARM Cortex-M4 in cycles. These implementations are in plain C and target 128 bits security.

Scheme	Keygen	Encaps	Decaps
ROLLO-I	16 927 603	1 926 332	7 009 943
ROLLO-II	85 063 257	6 844 408	17 321 266
RQC	5 756 747	11 340 541	71 551 978
frodokem640shake	91 940 068	109 310 982	109 009 172
kyber512	653 616	883 740	981 642
newhope512cca	715 680	1 128 510	1 186 054
ntruhps2048509	106 694 544	2 838 551	7 766 558
ntrulpr653	56 520 202	112 440 360	168 157 956
sikep434	672 303 199	1 100 796 989	1 174 307 957
sntrup653	599 438 684	56 563 524	170 044 505

Table 18. Performances of several KEM on ARM Cortex-M4 in cycles. These implementations features Cortex-M4 specific optimizations and target 128 bits security.

Scheme	Keygen	Encaps	Decaps
frodokem640aes	48 350 369	47 135 457	46 604 758
kyber512	470 998	596 970	555 224
newhope512cca	582 009	870 621	825 352
ntruhps2048509	77 457 221	606 804	555 866
sikep434	48 264 153	78 912 215	84 277 568

References

[AAB+17a] Melchor, C.A., et al.: Ouroboros-R. NIST Post-Quantum Cryptography Standardization Project (Round 1) (2017). https://pqc-ouroborosr.org

[AAB+17b] Melchor, C.A., et al.: Rank Quasi-Cyclic (RQC). NIST Post-Quantum Cryptography Standardization Project (Round 1) (2017). https://pqc-rqc.org

[AAB+18] Melchor, C.A., et al.: ROLLO - Rank-Ouroboros, LAKE & LOCKER. NIST Post-Quantum Cryptography Standardization Project (Round 1) (2018). https://pqc-rollo.org

[AAB+19a] Melchor, C.A., et al.: ROLLO - Rank-Ouroboros, LAKE & LOCKER. NIST Post-Quantum Cryptography Standardization Project (Round 2) (2019). https://pqc-rollo.org

[AAB+19b] Melchor, C.A., et al.: Rank Quasi-Cyclic (RQC). NIST Post-Quantum Cryptography Standardization Project (Round 2) (2019). https://pqc-rqc.org

[AAB+20a] Melchor, C.A., et al.: ROLLO - Rank-Ouroboros, LAKE & LOCKER. NIST Post-Quantum Cryptography Standardization Project (Round 2) (2020). https://pqc-rollo.org

[AAB+20b] Melchor, C.A., et al.: Rank Quasi-Cyclic (RQC). NIST Post-Quantum Cryptography Standardization Project (Round 2) (2020). https://pqc-rqc.org

[AB+] Aragon, N., Bidoux, L., et al.: RBC Library. Version 1.0. https://rbc-lib.org

[ABC+] Melchor, C.A., et al.: Constant-time algorithms for ROLLO

[ABC+19] Al Abdouli, A.S., Bellini, E., Caullery, F., Manzano, M., Mateu, V.: Rank-metric encryption on arm-cortex M0: porting code-based cryptography to lightweight devices. In: Proceedings of the 6th ASIA Public-Key Cryptography Workshop (2019)

[ABC+20] Albrecht, M.R., et al.: Classic McEliece. NIST Post-Quantum Cryptography Standardization Project (Round 3) (2020). https://classic.mceliece.org

[ABD+17a] Aragon, N., et al.: LAKE Low rAnk parity check codes Key Exchange. NIST Post-Quantum Cryptography Standardization Project (Round 1) (2017)

[ABD+17b] Aragon, N., et al.: LOCKER-LOw rank parity ChecK codes EncRyption. NIST Post-Quantum Cryptography Standardization Project (Round 1) (2017)

[ABG+19] Aragon, N., Blazy, O., Gaborit, P., Hauteville, A., Zémor, G.: Durandal: a rank metric based signature scheme. In: Ishai, Y., Rijmen, V. (eds.) EUROCRYPT 2019. LNCS, vol. 11478, pp. 728–758. Springer, Cham (2019). https://doi.org/10.1007/978-3-030-17659-4_25

[AGM+] Aranha, D.F., Gouvêa, C.P.L., Markmann, T., Wahby, R.S., Liao, K.: RELIC is an Efficient LIbrary for Cryptography. https://github.com/relic-toolkit/relic

[Ale03] Alekhnovich, M., et al.: More on average case vs approximation complexity. In: Proceedings of the 44th Annual IEEE Symposium on Foundations of Computer Science, pp. 298–307 (2003)

[ALH10] Aranha, D.F., López, J., Hankerson, D.: Efficient software implementation of binary field arithmetic using vector instruction sets. In: Abdalla, M., Barreto, P.S.L.M. (eds.) LATINCRYPT 2010. LNCS, vol. 6212, pp. 144–161. Springer, Heidelberg (2010). https://doi.org/10.1007/978-3-642-14712-8_9

[ALR18] Augot, D., Loidreau, P., Robert, G.: Generalized Gabidulin codes over fields of any characteristic. Des. Codes Crypt. 86(8), 1807–1848 (2018)

[AMAB+20a] Melchor, C.A., et al.: BIKE: Bit Flipping Key Encapsulation. NIST Post-Quantum Cryptography Standardization Project (Round 3) (2020). https://bikesuite.org

[AMAB+20b] Melchor, C.A., et al.: Hamming Quasi-Cyclic (HQC). NIST Post-Quantum Cryptography Standardization Project (Round 3) (2020). https://pqc-hqc.org

[AMADG21] Melchor, C.A., Aragon, N., Dyseryn, V., Gaborit, P.: Fast and secure key generation for low rank parity check codes cryptosystems. In: 2021 IEEE International Symposium on Information Theory (ISIT) (2021)

[BBGM19] Bettaieb, S., Bidoux, L., Gaborit, P., Marcatel, E.: Preventing timing attacks against RQC using constant time decoding of Gabidulin codes. In: Ding, J., Steinwandt, R. (eds.) PQCrypto 2019. LNCS, vol. 11505, pp. 371–386. Springer, Cham (2019). https://doi.org/10.1007/978-3-030-25510-7_20

[BMVT78] Berlekamp, E., McEliece, R., Van Tilborg, H.: On the inherent intractability of certain coding problems (corresp.). IEEE Trans. Inf. Theory 24(3), 384–386 (1978)

[Gab85] Gabidulin, E.M.: Theory of codes with maximum rank distance. Problemy Peredachi Informatsii 21(1), 3–16 (1985)

[GMRZ13] Gaborit, P., Murat, G., Ruatta, O., Zémor, G.: Low rank parity check codes and their application to cryptography. In: Proceedings of the Workshop on Coding and Cryptography (WCC) (2013)

[GT07] Gaudry, P., Thomé, E.: The MPFQ library and implementing curve-based key exchanges (2007)

[GT08] Gaudry, P., Thomé, E.: MPFQ, a finite field library (2008). https://mpfq.gitlabpages.inria.fr/

[Gue10] Gueron, S.: Intel Advanced Encryption Standard (AES) new instructions set (2010)

[HHK17] Hofheinz, D., Hövelmanns, K., Kiltz, E.: A modular analysis of the Fujisaki-Okamoto transformation. In: Kalai, Y., Reyzin, L. (eds.) TCC 2017. LNCS, vol. 10677, pp. 341–371. Springer, Cham (2017). https://doi.org/10.1007/978-3-319-70500-2_12

[HMV06] Hankerson, D., Menezes, A.J., Vanstone, S.: Guide to Elliptic Curve Cryptography. Springer, Heidelberg (2006)

[KRSS20] Kannwischer, M.J., Rijneveld, J., Schwabe, P., Stoffelen, K.: PQM4: Post-quantum crypto library for the ARM Cortex-M4 (2020). https://github.com/mupq/pqm4

[LMB+19] Lablanche, J., Mortajine, L., Benchaalal, O., Cayrel, P.-L., El Mrabet, N.: Optimized implementation of the NIST PQC submission ROLLO on microcontroller. IACR Cryptology ePrint Archive 2019:787 (2019)

[Loi05] Loidreau, P.: A Welch–Berlekamp like algorithm for decoding Gabidulin codes. In: Ytrehus, Ø. (ed.) WCC 2005. LNCS, vol. 3969, pp. 36–45. Springer, Heidelberg (2006). https://doi.org/10.1007/11779360_4

[McE78] McEliece, R.J.: A public-key cryptosystem based on algebraic coding theory, NASA (1978)

[Min] Minunit, a minimal unit testing framework for C/C++. https://github.com/siu/minunit

[Nis16] NIST Post-Quantum Standardization Process (2016). https://csrc.nist.gov/Projects/Post-Quantum-Cryptography

[Ope] OpenSSL. https://www.openssl.org/

[Ore33] Ore, O.: On a special class of polynomials. Trans. Am. Math. Soc. 35(3), 559–584 (1933)

[Por16] Pornin, T.: BearSSL: A smaller SSL/TLS library (2016). https://bearssl.org/

[PQC] PQClean: Clean, portable, tested implementations of post-quantum cryptography. https://github.com/PQClean/PQClean

[S+01] Shoup, V., et al.: NTL: a library for doing number theory (2001). https://www.shoup.net/ntl/

[Sup] SUPERCOP, measuring the performance of cryptographic software. https://bench.cr.yp.to/supercop.html

Security Analysis of a Cryptosystem Based on Subspace Subcodes

Thierry P. Berger[1], Anta Niane Gueye[2], Cheikh Thiecoumba Gueye[2],
M. Anwarul Hasan[3], Jean Belo Klamti[3], Edoardo Persichetti[4],
Tovohery H. Randrianarisoa[4(✉)], and Olivier Ruatta[1]

[1] Université de Limoges, XLIM, UMR 7252, Limoges, France
{thierry.berger,olivier.ruatta}@unilim.fr
[2] Faculté des Sciences et Techniques, Universit Cheikh Anta Diop, Dakar, Senegal
antaniane@orange.sn, cheikht.gueye@ucad.edu.sn
[3] Department of Electrical and Computer Engineering, University of Waterloo,
Waterloo, Canada
{ahasan,jbklamti}@uwaterloo.ca
[4] Department of Mathematical Sciences, Florida Atlantic University,
Boca Raton, USA
{epersichetti,trandrianarisoa}@fau.edu

Abstract. In 2019, Berger et al. introduced a code-based cryptosystem using quasi-cyclic generalized subspace subcodes of Generalized Reed-Solomon codes (GRS). In their scheme, the underlying GRS code is not secret but a transformation of codes over $\mathbb{F}_{2^m}$ to codes over $\mathbb{F}_2$ is done by choosing some arbitrary $\mathbb{F}_2$-subspaces V_i of $\mathbb{F}_{2^m}$ and by using the natural injection $V_i \subset \mathbb{F}_{2^m} \hookrightarrow \mathbb{F}_2^m$. In this work, we study the security of the cryptosystem with some additional assumption. In addition to the knowledge of the GRS code, we introduce a new kind of attack in which the subspaces are corrupted. We call this attack "known subspace attack" (KSA). Although this assumption is unrealistic, it allows us to better understand the security of this scheme. We are able to show that the original parameters are not secure; in practice this however does not break the original proposal. In this paper, we provide new parameters for Berger et al.'s scheme which are secure against the known subspace attack.

Keywords: Coding theory · subspace subcode · generalized subspace subcode · code-based cryptography

1 Introduction

With quantum computers getting ever closer to realization, several public-key cryptographic schemes need to be replaced by quantum-resistant (or *post-quantum*) alternatives. To address this issue, NIST launched a post-quantum

This work is partially funded by the National Science Foundation (NSF) Grant CNS-1906360 and by the Ripple Impact Fund/Silicon Valley Community Foundation Grant 2018-188473.

A. Wachter-Zeh et al. (Eds.): CBCrypto 2021, LNCS 13150, pp. 42–59, 2022.
https://doi.org/10.1007/978-3-030-98365-9_3

cryptography standardization process with a call for submissions in 2016. The process is now in the final phase [32], with four candidate algorithms for Public Key Encryption (PKE) and Key Encapsulation Mechanism (KEM); two of the candidates are lattice-based and the other two are code-based and isogeny-based.

The code-based system is largely based on R. J. McEliece's public-key scheme introduced in 1978 [28]. McEliece's scheme was the first public-key encryption scheme for which security relies on the hardness of a problem coming from coding theory, namely, the problem of decoding a random linear code without any visible structure. The general problem of maximum-likelihood decoding is known to be NP-complete [27]. Also, the problem of decoding up to half the minimum distance is considered to be hard, and it is not, at the current state of the art, subject to a dedicated quantum algorithm (such as the case of factoring or discrete logarithm [37]). In fact, the only quantum speed up is given by applying Grover's algorithm, which would roughly cut in half the exponent in the complexity of a generic search algorithm, such as the information set decoding (ISD) algorithm. The main drawback of this cryptosystem is still represented by the size of the public key, which is too large for some applications. In McEliece's original proposal, the system used binary Goppa codes and the key size was 32.7 kB for a classical security of 65 bits [14]. In its modern declination, i.e. the code-based candidate algorithm called Classic McEliece, authors are still using binary Goppa codes, for conservative reasons which strongly relate to security and confidence. With the parameters chosen in the third round for NIST's post-quantum cryptography standardization, the public key size is approximately 261 kB for classical security of 128 bits [14].

In order to reduce the key size, a long track of research works has been devoted to variants of the McEliece cryptosystem. The first attempt, based on another paradigm than that of McEliece, was given by Niederreiter [31]. This scheme is considered a "dual" version, in the sense that it is equivalent to the McEliece cryptosystem from a security point of view and it also uses a different class of code (GRS codes). However, the use of GRS codes was proved to be unsecure [38]. Other ideas include replacing the underlying code family (e.g. with subcodes of generalized Reed-Solomon (GRS) codes), masking the underlying structure of the public code, or using codes with non-trivial automorphism groups [2,6,7,13,19,29,30,35]. However, most of these variants have shown insecurities, that have been exploited via structural attacks [4,15–17,39]. Rank metric codes were also suggested [19,21]. Although the scheme of [19], and further variations [20,26,36] were shown to be insecure [23,34], the scheme of [21] which uses Low rank parity check codes (LRPC) is promising and in fact a scheme based on LRPC codes was submited to the Round 2 of the NIST standardization process [33].

Recently, in 2019, two new proposals were introduced by Berger et al. [12] and Khathuria et al. [25], respectively. The former suggests to use Quasi-Cyclic (QC) subspace subcodes of Reed-Solomon codes, while the latter is based instead on expanded Reed-Solomon Codes. Unfortunately, the scheme of Khathuria et al. was subsequently cryptanalyzed by Couvreur and Lequesne in [14].

Our Contribution and Paper Organization

We introduce a new security model, which we call Known Subspace Attack (KSA). In this model, the adversary is given the knowledge of the secret subspace used in the scheme of Berger et al. [12] which we denote in this paper as the BGKR (Berger-Gueye-Klamti-Ruatta) scheme. The KSA assumption helps us to better understand the intrinsic security of the cryptosystem. Furthermore, it opens the door to new scenarios; for instance, disclosing part of the private key could help greatly reduce its size. We present a detailed security analysis of the cryptosystem in this model, and show that the original scheme parameters are not secure in this case. We then devise new parameters that show how it is possible to resist the attack and secure the cryptosystem, in the case that attacks in the KSA model ever became a reality.

The paper is organized as follows. We begin in Sect. 2 by briefly defining notation, and summarizing some coding theory notions that are relevant to our work. In Sect. 3, we recall the BGKR scheme. We first present a generic description that uses non-QC codes, and then describe the QC version of [12]. We then proceed to introduce the KSA security model (Sect. 4) and give a security analysis of the scheme in this model (Sect. 5). Finally, we conclude in Sect. 6.

2   Preliminaries

To begin, we fix the following notation conventions. We write $\mathbb{F}_2$ for the binary finite field, and $\mathbb{F}_{2^m}$ for an extension of $\mathbb{F}_2$ of degree m. We indicate with $\mathsf{GL}_2(m)$ the General Linear group of $m \times m$ invertible matrices over $\mathbb{F}_2$, and with $\mathbb{E} = \mathbb{F}_2^m$ the set of binary vectors of length m. We denote with $\mathrm{Diag}(L)$ the diagonal matrix with the vector L on the diagonal and with $\otimes$ the Kronecker product of two matrices. Finally, we write $SS_V(\mathcal{C})$ for a subspace subcode of a code $\mathcal{C}$ relative to the subspace V of $\mathbb{F}_{2^m}$, and $SS_{\overline{V}}(\mathcal{C})$ for a generalized subspace subcode of a code $\mathcal{C}$ relative to $\overline{V} = V_1 \times ... \times V_n \subset \mathbb{F}_{2^m}^n$, where n is the length of $\mathcal{C}$ and $V_1,...,V_n$ are subspaces of $\mathbb{F}_{2^m}$ of dimension μ.

We now give some standard coding theory definitions that are relevant to our work.

2.1   GRS Codes

Definition 1. *Let $x = (x_0, x_1, ..., x_{n-1}) \in \mathbb{F}_{2^m}^n$ be a vector with pairwise distinct entries. Let $y = (y_0, y_1, ..., y_{n-1}) \in \mathbb{F}_{2^m}^n$ be a vector whose entries are all nonzero. The Generalized Reed-Solomon (GRS) code with support x and multiplier y of dimension k is the set denoted by $\mathrm{GRS}_k(x, y)$ and defined by*

$$\mathrm{GRS}_k(x, y) = \{(y_0 f(x_0), ..., y_{n-1} f(x_{n-1})) \ s.t. \ f \in \mathbb{F}_{2^m}[x], \ deg(f) < k\}$$

When $y = (1, ..., 1)$, then the code $\mathrm{GRS}_k(x, y)$ is said to be a *Reed-Solomon* code and is simply denoted as $\mathrm{RS}_k(x)$.

2.2   Shortened Codes and Punctured Codes

Let C be an $\mathbb{F}_{2^m}$-linear code of length n and I a subset of $\{0, ..., n-1\}$. Let $\tilde{C}_I$ be the set defined as $\tilde{C}_I = \{c = (c_1, ..., c_n) \in C \text{ s.t. } c_i = 0, \forall i \in I\}$.

Definition 2. *The punctured code of C on positions I is the code $\mathrm{Punct}_I(C)$ obtained from the codewords of C by deleting the coordinates indexed by I. The shortened code of C on positions I is the code $\mathrm{Short}_I(C)$ obtained by puncturing its subcode $\tilde{C}_I$ on I, i.e. $\mathrm{Short}_I(C) = \mathrm{Punct}_I(\tilde{C}_I)$.*

The following theorem gives us the link between punctured and shortened codes, and some properties on puncturing and shortening of codes.

Theorem 1. *([24] see p. 17 Theorem 1.5.7) Let C be an $[n, k, d]$ $\mathbb{F}_{2^m}$-linear code. Let I a subset of $\{0, 1, ..., n-1\}$ such that $|I| = r$. Then:*

1. $\mathrm{Short}_I(C^\perp) = (\mathrm{Punct}_I(C))^\perp$
2. *If $r < d$, then $\mathrm{Short}_I(C^\perp)$ and $\mathrm{Punct}_I(C)$ have dimensions $n - t - k$ and k, respectively;*
3. *If $r = d$ and I is the set of coordinates where a minimum weight codeword is nonzero, then $\mathrm{Short}_I(C^\perp)$ and $\mathrm{Punct}_I(C)$ have dimensions $n - d - k + 1$ and $k - 1$, respectively;*

2.3   Binary m-Block Codes

We now recall some notions of m-block codes and of binary image of an $\mathbb{F}_{2^m}$-linear code C. For more details, see [8, 9, 11].

Definition 3. *Let $(A, +)$ be an additive group. An additive code of length n over A is an additive subgroup of $(A^n, +)$.*

Definition 4. *A binary m-block code of length n over $\mathbb{E} = \mathbb{F}_2^m$ is an additive code over the additive group $(\mathbb{E}, +)$, i.e. a subgroup of $(\mathbb{E}^n, +)$. The integer m is the block size.*

Let $\mathcal{B} = \{b_1, ..., b_m\}$ be a basis of $\mathbb{F}_{2^m}$ over $\mathbb{F}_2$. We denote by $\phi_\mathcal{B}$ the corresponding $\mathbb{F}_2$-linear isomorphism $\mathbb{F}_{2^m} \mapsto \mathbb{F}_2^m$ and by $\Phi_\mathcal{B}$ the extension of $\phi_\mathcal{B}$ to the whole space $\mathbb{F}_{2^m}^n$ i.e. for all $x = \sum_{i=0}^{m-1} \alpha_i b_i \in \mathbb{F}_{2^m}$ with $\alpha_i \in \{0, 1\}$ and $c = (c_0, c_1, ..., c_{n-1}) \in \mathbb{F}_{2^m}^n$ we have

$$\phi_\mathcal{B}(x) = (\alpha_0, \alpha_1, ..., \alpha_{m-1}) \in \mathbb{F}_2^m \text{ and } \Phi_\mathcal{B}(c) = (\phi_\mathcal{B}(c_0), ..., \phi_\mathcal{B}(c_{n-1})) \in \mathbb{F}_2^{nm}.$$

For the construction of binary m-block codes from $\mathbb{F}_{2^m}$-linear codes of length n C we should fix a basis $\mathcal{B} = \{b_1, ..., b_m\}$ of $\mathbb{F}_{2^m}$ over $\mathbb{F}_2$. Then, the image $\Phi_\mathcal{B}(C) = \{\Phi_\mathcal{B}(c) = (\phi_\mathcal{B}(c_0), ..., \phi_\mathcal{B}(c_{n-1})) \text{ s.t. } c \in C\}$ of C by $\Phi_\mathcal{B}$ is a binary m-block code of length n.

Definition 5. *Let C be an $[n, k, d]$ $\mathbb{F}_{2^m}$-linear code. Let $\mathcal{B}$ be a basis of $\mathbb{F}_{2^m}$ over $\mathbb{F}_2$. Then, the binary image of C relative to the basis $\mathcal{B}$ is $Im_{bin}(C) = \Phi_\mathcal{B}(C)$.*

The parameters of a binary image of an $[n, k, d]$ $\mathbb{F}_{2^m}$-linear code $\mathcal{C}$ seen as a binary m-block code are the same than the original $\mathbb{F}_{2^m}$-linear code. For the construction of a generator matrix G_{bin} of the binary image $Im_{bin}(\mathcal{C})$ of a code $\mathcal{C}$ from a generator matrix G of $\mathcal{C}$ see [11, Proposition 4].

In order to introduce the equivalence of m-block codes we need the following definitions.

Definition 6. *An isometry for an m-block code is an $\mathbb{F}_2$-linear map on $\mathbb{E}^n$ which preserves the m-block distance.*

Definition 7. *Let $x = (\overline{x}_1, ..., \overline{x}_n) \in \mathbb{E}^n = \mathbb{F}_2^{nm}$, so that, for all $i = 1, ..., n$, we have $\overline{x}_i = (x_{i,1}, x_{i,2}, ..., x_{i,m}) \in \mathbb{F}_2^m$. We call multiplier the map defined by*

$$L(x) = \mathrm{Diag}(L) \cdot x^T$$

where $L = (\Lambda_1, ..., \Lambda_n) \in \mathsf{GL}_2(m)^n$ is a sequence of elements of $\mathsf{GL}_2(m)$.

We can see that a multiplier is an $\mathbb{F}_2$-isometry of $\mathbb{E}^n$. Then $\mathbb{F}_2$-isometries of $\mathbb{E}^n$ can be characterized by the following theorem.

Theorem 2 (Theorem 1 of [11]). *The $\mathbb{F}_2$-isometries of $\mathbb{E}^n$ (i.e. linear isomorphisms preserving the Hamming block-weight) form a group generated by the m-block permutations and their multipliers.*

Next, we recall different definition of m-block codes equivalence.

Definition 8. *Let $\mathcal{C}$ and $\mathcal{C}'$ be two m-block codes of length n over $\mathbb{E}$.*

  - *$\mathcal{C}$ and $\mathcal{C}'$ are equivalent if there is an isometry $f = L \circ \pi$, (where L is a multiplier and π a permutation) such that $\mathcal{C}' = f(\mathcal{C})$.*
  - *$\mathcal{C}$ and $\mathcal{C}'$ are equivalent by permutation if there is a permutation at block level $\pi \in \mathrm{Sym}(n)$ such that $\mathcal{C}' = \pi(\mathcal{C})$.*
  - *$\mathcal{C}$ and $\mathcal{C}'$ are equivalent by multiplier if there is a multiplier $L \in \mathsf{GL}_2(m)^n$ such that $\mathcal{C}' = L(\mathcal{C})$.*

2.4 Subspace Subcodes

Definition 9. *Let $\mathcal{C}$ be a binary m-block code of length n and V be an $\mathbb{F}_2$-subspace of $\mathbb{F}_{2^m}$ of dimension $\mu \leq m$. The Subspace Subcode (SS) over V of $\mathcal{C}$ is the $\mathbb{F}_2$-linear code $\mathcal{C}_{|V^n} = \mathcal{C} \cap V^n$.*

To obtain an effective representation of the code $\mathcal{C}_{|V^n}$, it is necessary to choose a basis of V as a subspace of $\mathbb{F}_2^m$ of dimension μ. It can be done for instance by choosing a generator matrix M_V of V as a linear code of length m. The elements of V are replaced by their μ coordinates on this basis. By this identification, we obtain a μ-block code of length n, denoted $\mathrm{SS}_V(C)$. For the rest of this work, a μ-subspace subcode will canonically be associated to its μ-block representation $\mathrm{SS}_V(C)$. Note that the notion of subspace subcodes of m-block codes can be naturally extended to linear codes over $\mathbb{F}_2$ by constructing first a binary image of these codes.

Definition 10. *Let ψ be an $\mathbb{F}_2$-projection of rank μ of $\mathbb{F}_{2^m}$ onto an $\mathbb{F}_2$-subspace $V \subset \mathbb{F}_{2^m}$ of dimension μ. Let Ψ be the action of ψ on each coordinate of a word in $\mathbb{F}_{2^m}^n$. Let $\mathcal{C}$ be an $\mathbb{F}_{2^m}$-linear. The (μ-)projected code relative to ψ of $\mathcal{C}$ is the code $\mathrm{P}_\psi(\mathcal{C}) = \Psi(\mathcal{C})$.*

The link between subspace subcodes and projected codes as μ-block codes of length n is described by the following proposition.

Proposition 1 (Proposition 9 of [11]). *The μ-projected codes are the duals of μ-subspace subcodes. More precisely, the dual of the μ-block subcode $\mathrm{SS}_V(\mathcal{C})$ relative to a generator matrix M_V is the μ projected code $\Psi(\mathcal{C}^\perp)$, where the $m \times \mu$ projection matrix of π is M_V^T.*

For more details see [11].

2.5 Generalized Subspace Subcodes

Definition 11. *Let $\overline{V} = \prod_{i=1}^n V_i$ be the cartesian product of $\mathbb{F}_2$-subspaces of $\mathbb{F}_{2^m}$ of dimension μ and $\mathcal{C}$ be a binary μ-block code of length n. The* Generalized Subspace Subcode (GSS) *of $\mathcal{C}$ relative to $\overline{V}$ is $\mathrm{GSS}_{\overline{V}}(\mathcal{C}) = \mathcal{C} \cap \overline{V}$.*

For short, we will use the acronym GSS-GRS to indicate a Generalized Subspace Subcode of a GRS code.

The μ-block representation of $\mathrm{GSS}_{\overline{V}}(\mathcal{C})$ is constructed as before by choosing a basis for each μ-subspace V_i. An equivalent characterization for μ-subspace subcodes of a code $\mathcal{C}$ can be then given as follows. The generalized subspace subcodes of an m-block code $\mathcal{C}$ are the μ-subspace subcodes of the m-block codes that are multiplier-equivalent to $\mathcal{C}$. Note that, if our code $\mathcal{C}$ is a binary image of an $\mathbb{F}_{2^m}$-linear code, the notion of multiplier equivalence at m-block level is more general than those at $\mathbb{F}_{2^m}$ level.

3   Scheme Description

In this section, we recall the description of the BGKR scheme, or, more precisely, of its key generation algorithm. Note that the scheme introduced in [12] is based on quasi-cyclic subspace subcodes of GRS codes; however, in order to have a better understanding of the construction and its security features, we first describe a generic version, by which we mean a scheme that uses non-quasi-cyclic codes.

3.1   Description of Generic Scheme

We begin by describing this version, so that we can focus first on the main difference, compared with traditional schemes such as Classic McEliece, which is the construction of the public code. Note that this version is not suitable in practice, due to the large size of the public key. The goal, thus, is just to explain the generalized subspace construction so that the reader can get accustomed to it. In the next section, we will focus on the security analysis of the scheme actually proposed in [12].

Rationale. A crucial point in public key code-based cryptography is to hide the structure of the secret code. In [12], however, the key generation is based on a GRS code which, by construction, does not need to be hidden. Instead, the scheme uses the notion of GSS, that allows to destroy the original GRS structure. This is because, unlike traditional schemes that use subfield subcodes, the BGKR scheme uses subspace subcodes, in conjunction with a block-puncturing technique, which is a block projection on vector subspaces. In the SS case, the projections are the same on all blocks, while in the Generalized case, the projections are different.

For the key generation, the following items are required and must be public:

- The extension degree m of the finite field $\mathbb{F}_{2^m}$.
- Parameters $[n, k, d = n - k + 1]$ of a GRS code.
- The dimension $\mu < m$ of the subspaces V_i of $\mathbb{E} = \mathbb{F}_2^m$.
- A random public GRS code $\mathcal{C}$ of length n and dimension k over $\mathbb{F}_{2^m}$.
- A binary image $\mathcal{C}_{bin}$ of $\mathcal{C}$ defined by its generator matrix G_{bin} of size $mk \times mn$ or by its parity check matrix H_{bin}. The basis $\mathcal{B}$ used for computing this binary image is not a secret and could be published (Fig. 1).

The algorithm's description is summarized below.

Algorithm 1 *Input:*

  · *The extension degree m of the finite field $\mathbb{F}_{2^m}$.*
  · *Parameters $[n, k, d = n - k + 1]$ of a GRS code.*
  · *The dimension $\mu < m$ of subspaces V_i of $\mathbb{E}$.*

---

1. *Choose randomly a GRS code $\mathcal{C}$ of length n and dimension k.*
2. *Construct the binary image $\mathcal{C}_{bin}$ of $\mathcal{C}$ of parameters $[mn, mk]$. $\mathcal{C}_{bin}$ is defined by its generator matrix G_{bin} of size $mk \times mn$ or its parity check matrix H_{bin}.*
3. *Choose randomly n binary matrices M_i of size $m \times \mu$ and full rank μ. For each $i = 0, 1, ..., n - 1$, M_i^T is the generator matrix of a subspace V_i in the subspace subcodes construction.*
4. *Compute $H_{gss} = H_{bin}\Psi$ where $\Psi = \mathrm{Diag}(M_0, M_1, ..., M_{n-1})$. H_{gss} is the parity-check matrix of the GSS-GRS $\mathcal{C}_{gss}$ over subspaces V_i.*
5. *Choose randomly a permutation $\Pi \in \mathrm{Sym}(N)$ defined by a matrix $\Pi \in \mathbb{F}_2^{N \times N}$ which acts on $\mathbb{F}_2^N$ where $N = n\mu$.*
6. *Construct the GSS-GRS $\mathcal{C}'_{gss}$ defined by its generator matrix $G'_{gss} = G_{gss}\Pi$.*
7. *Construct a subcode $\mathcal{C}_{pub}$ of $\mathcal{C}'_{gss}$ of dimension k_{pub} and generator matrix $G_{pub} = [I_{k_{pub}}|R]$.*
8. *The public key is $pk = R$.*
9. *The private key sk is composed by the n matrices M_i and the permutation Π.*

Fig. 1. Key Generation of the Generic Scheme

Note that the m-block distance of $\mathcal{C}_{bin}$ is equal to $d = n - k + 1$ but its Hamming distance is greater than or equal to d. Then, it is possible to decode up to $t = \lfloor (d-1)/2 \rfloor$ errors with $\mathcal{C}_{bin}$ as an m-block code (and then at least t in Hamming distance).

In the above algorithm, the binary code $\mathcal{C}'_{gss}$ is equivalent to $\mathcal{C}_{gss}$ and the public code $\mathcal{C}_{pub}$ is a subcode of $\mathcal{C}'_{gss}$. The public code $\mathcal{C}_{pub}$ is an $[N, k_{pub}]$ binary code with $N = n\mu$. The minimal distance of $\mathcal{C}$ corresponds to the minimal distance of a random binary linear code, and it is thus lower-bounded by $n - k + 1$. However, the decoding algorithm derived from the GRS code can correct until t errors. Note also that the error correction capacity of the decoding algorithm in this case is lower than the actual capacity of the public code.

3.2   Description of the BGKR Scheme

The key generation in this second version is identical to the previous one, except for the following differences. First, in addition to the previous parameters, the scheme features the quasi-cyclic order ℓ and the index s for the GRS code. Then, the public GRS code is quasi-cyclic and not Reed-Solomon. Finally, the projection in the construction of the generalized subspace subcode needs to preserve the quasi-cyclicity, and therefore is same on each orbit.

Before explicitly defining the key generation algorithm, it is important to recall the two different definitions of quasi-cyclic permutations, and then the two different representations of quasi-cyclic codes.

- Type 1, orbit order: the quasi-cyclic permutation acts as a simultaneous application of the shift permutation on each block of size ℓ. It is then possible to construct a generator matrix of a quasi-cyclic code which is of ℓ-block circulant matrices.
- Type 2, shift order: the quasi-cyclic permutation is $shift^s$, and corresponds to a permutation on blocks of size s.

A quasi-cyclic code of Type 1 can be transformed to a quasi-cyclic code of Type 2 via a permutation matrix P_s. In case the public code is quasi-cyclic of Type 1, then we can just use the transformation P_s to transform the problem with codes of Type 2. Hence, we will assume that the public and private codes are quasi-cyclic of Type 2. In the construction of a quasi-cyclic public key, care must be taken to use transformations that are compatible with quasi-cyclicity. We start with a public QC-GRS code $\mathcal{C}$ of order ℓ and index s of Type 2. The binary image of this code $\mathcal{C}$ is also quasi-cyclic of Type 2. However, if we want to obtain a generalized subspace subcode, we have to apply the projection matrices on this binary image.

In the second part of the key generation algorithm, we must apply a permutation on G_{gss} which destroys the μ-block structure but preserves the Type 2 quasi-cyclic structure at binary level. Such a permutation Π is of the form $\Pi = \Pi_1 \Pi_2$ where

- the matrix Π_1 of size $N \times N$ is given by: $\Pi_1 = P_\pi \otimes I_\ell$ and P_π is the matrix of a permutation $\pi \in \text{Sym}(s\mu)$,
- the matrix Π_2 is a matrix of size $N \times N$ and it is given by the matrix $\Pi_2 = \text{Diag}(P_{shift}^{i_0}, P_{shift}^{i_1}, ..., P_{shift}^{i_{s'-1}})$ where $s' = s\mu$ and P_{shift} is a matrix of a circular permutation of size ℓ. Integers $i_0, i_1, ..., i_{s',-1}$ are chosen randomly from $\{1, 2, ..., \ell\}$ (Fig. 2).

Algorithm 2 *Input:*

- *The extension degree m of the finite field $\mathbb{F}_{2^m}$.*
- *The dimension $\mu < m$ of subspaces V_i of $\mathbb{E}$.*
- *Integers s, and ℓ such that the parameters of a GRS code is given by $[n = s\ell, k, d = n - k + 1]$.*
- *The length $N = s\ell\mu$ of quasi-cyclic public code*

---

1. *Choose a QC-GRS code $\mathcal{C}$ (of Type 2) and compute the parity check matrix H_{bin} of the binary image $\mathcal{C}_{bin}$ of the QC-GRS code $\mathcal{C}$.*
2. *Choose randomly s binary matrices $M_i \in \mathbb{F}_2^{m \times \mu}$ of full rank μ with $i = 0, 1, ..., s-1$ and set $\Psi = \text{diag}(M_0, ...M_{s-1}, M_0, ..., M_{s-1}, ..., M_0, ..., M_{s-1})$.*
3. *Compute generator matrix G_{gss} of the GSS-GRS code $\mathcal{C}_{gss}$ from the code $\mathcal{C}_{bin}$ such that its parity-check is given by $H_{gss} = H_{bin}\Psi$.*
4. *Reorder G_{gss} to obtain a quasi-cyclic code of Type I, index ℓ and order s'.*
5. *Choose random a permutation $\Pi = \Pi_1\Pi_2 \in \text{Sym}(s\ell\mu)$ compatible with the binary quasi-cyclic structure.*
6. *Computer the equivalent code $\mathcal{C}'_{gss}$ of $\mathcal{C}_{gss}$ of generator matrix $G'_{gss} = G_{gss}\Pi$.*
7. *The public code $\mathcal{C}_{pub}$ is generated quasi-cyclically by one or more codewords of $\mathcal{C}'_{gss}$ chosen randomly.*
8. *The private key is composed of the s matrices M_i and the permutation Π.*

Fig. 2. Key Generation of the BGKR Scheme

Remark 1. Since the underlying Generalized Reed-Solomon code is not secret, the addition of the quasi-cyclic structure is not harmful. In fact, traditional threats such as folding attacks or square code attacks for distinguishing the underlying code can't be used anymore.

4   New Security Model

We now proceed to model attacks in which we assume the adversary has the knowledge of the secret subspace, which we therefore call *Known Subspace Attacks (KSA)*. To begin, we summarize both the public and the private objects in the design of the BGKR scheme. In the generic, non-QC scheme, the public data consists of the GRS code $\mathcal{C}$, its binary image $\mathcal{C}_{bin}$ described by the parity check matrix H_{bin}, the code $\mathcal{C}_{pub}$ and its generator matrix G_{pub}. The knowledge of the GRS code $\mathcal{C}$ and its binary image does not affect the security of the scheme. On the other hand, the private objects are the generator matrices M_i of subspaces V_i used for the construction of the GSS $\mathcal{C}_{gss}$. This later code has parity check matrix H_{gss}. Another secret object is the permutation Π used to obtain the code $\mathcal{C}'_{gss}$, which is equivalent to $\mathcal{C}_{gss}$. Note that the code $\mathcal{C}_{gss}$ itself is also kept hidden. According to the construction of GSS-GRS, the knowledge of subspaces V_i implies the knowledge of $\mathcal{C}_{gss}$. These private and public objects satisfy the following equation.

$$H_{bin}\Psi\Pi G_{pub}^{T} = 0. \tag{1}$$

In Eq. (1), the unknowns are the coefficients of the matrix Ψ, and they can lead to the construction of the matrices M_i. In our security analysis we use the following strong hypothesis:

Hypothesis 1. *The attacker knows the GSS-GRS code $\mathcal{C}_{gss}$.*

Using this hypothesis we will be able to clarify on which assumption the security of BGKR scheme relies. We have the following definition:

Definition 12. *The attack model in which the attacker knows the generator matrix of the GSS-GRS code $\mathcal{C}_{gss}$ is called Known Subspace Attack (KSA).*

Under the KSA Hypothesis, Ψ is known and therefore we know $H_{gss} = H_{bin}\Psi$. Therefore, Eq. (1) becomes:

$$H_{gss}\Pi G_{pub}^{T} = 0. \tag{2}$$

Remark 2. When the GRS code $\mathcal{C}$ is secret, the attack consists of finding the GRS code $\mathcal{C}$ from the GSS-GRS code $\mathcal{C}_{gss}$ and then the corresponding scheme is an SSRS based cryptosystem. This problem was solved for $\mu \geq m/2$ by Lequesne and Couvreur in [14]. In the case of KSA, with the parameters originally proposed in [12], the GRS code or an equivalent instance can be rebuilt. That is why we assume that the GRS code is not secret.

Under Hypothesis 1, the cryptanalysis of the generic scheme corresponds to finding the permutation Π from the public code $\mathcal{C}_{pub}$ and the GSS-GRS code $\mathcal{C}_{gss}$. The recovering of Π, is then an instance of the computational version of the Equivalent Subcode (ES) problem, which we recall below.

Problem 1. (Equivalent Subcode (ES)).

Given two linear codes $\mathcal{C}$ and $\mathcal{D}$ of length of n and respective dimension k and k', $k' \leq k$, over the same finite field $\mathbb{F}_{q^m}$, is there a permutation σ of the support such that $\sigma(\mathcal{C})$ is a subcode of $\mathcal{D}$?

The ES problem was proved to be NP-complete for the generic case by Berger et al. in [10]. However, we do not know how hard is the instance in which the main code is a GSS-GRS code. In our security analysis, we assume that such an instance behaves similarly to a generic instance of the ES problem. We can then see that the natural way to get the permutation Π is to perform an exhaustive search or to use an algebraic approach.

5   Security Analysis of BGKR in the KSA Model

To begin, note we can actually simplify Eq. (2). Recall, in fact, that our equation is of the form $H_{gss}\Pi G_{pub}^T = 0$. Now, the rows of the generator matrix G_{pub} consist of quasi-cyclic shifts of the generator vectors. From our construction, one can see that the row spaces of $H_{gss}\Pi$ and G_{pub} are invariant under a common quasi-cyclic permutation, say Q. Hence if $H_{gss}\Pi v^T = 0$, then we also have $H_{gss}\Pi(vQ)^T = 0$ and vice versa. Therefore, Eq. (2) contains redundancies, and can be simplified as

$$H_{gss}\Pi G^T = 0, \tag{3}$$

where G is made of the vectors which generate the matrix G_{pub} cyclically. For the parameters proposed in [12], the public code is given by a matrix G which is made of a single vector. Therefore, when modelling our attacks, we will focus on solving Eq. (3).

5.1   Exhaustive Search

In the KSA model this type of attack is very simple as it consists of guessing the permutation Π and then checking if it satisfies Eq. (3). In the generic version of the scheme, Π is a standard permutation of size $N \times N$; since there are $N!$ choices for the permutation, the complexity of this attack is given by $\mathcal{O}(N!)$. It follows that properly chosen parameters could avoid a such attack. However, this is different in the case of the BGKR scheme, because the permutation Π is not only in quasi-cyclic form, but also a product of two quasi-cyclic matrices. We can use this notion to reduce the complexity of the exhaustive search attack in this case. The simplest way to do this is to guess the permutation matrices Π_1 and Π_2. In the end, the complexity of this attack corresponds essentially to $\mathcal{O}(s'!\ell^{s'-1})$, where $s' = s\mu$ is the index of quasi-cyclicity and ℓ is the order.

We now show how to apply the work of Georgiades [22] to improve on the basic exhaustive search. For this, we assume that the public code is generated by a single vector, hence we set $G = v$. The case of many generators is similar and does not change the overall complexity of the attack. Assume that the matrix $H_{gss} \in \mathbb{F}_2^{r \times N}$ is in the form:

$$H_{gss} = (a_{ij})_{1 \leq i \leq r, 1 \leq j \leq N} = [A|I]. \tag{4}$$

We want to solve Eq. (3) for Π. Through the form of H_{gss} given by (4), the solutions of the equation $H_{gss}x^T = 0$ are of the form

$$x = (f_1, \ldots, f_r, \lambda_1, \ldots, \lambda_{N-r}), \tag{5}$$

where $f_j = \sum_{i=1}^{N-r} u_{i,j} \lambda_i$, $j \in \{1, \ldots, r\}$, u_{ij} are fixed scalars and $\lambda_i \in \mathbb{F}_2$. Hence, instead of guessing the full permutation Π, it is enough to guess the last $N - r$ elements of $v\Pi^T$ among those of v, compute the remaining elements through the form of the solution of (4), and check if that solution is indeed a permutation of v. If there was no restriction on Π, the complexity of guessing the full permutation is $\mathcal{O}(N!)$ whereas with Georgiades' technique we have $\mathcal{O}\left(\frac{N!}{r!}\right)$. In the end, we obtain an improvement on the exhaustive search when the matrix Π is a standard permutation of size $N \times N$.

Coming back to our setting, we have a particular type of permutation matrix $\Pi = \Pi_1 \Pi_2$. Instead of guessing $N - r$ entries, we have to guess the blocks which cover these $N - r$ entries, that will amount to guessing at least $\lceil \frac{N-r}{\ell} \rceil$ positions for the permutation Π_1. Then, for the part coming from Π_2, for each block, we have to guess the shift permutation used in each block. By normalization, one of the permutation is the identity matrix. In total the complexity is given by

$$\mathcal{O}\left(\frac{(s'!)}{(s' - \lceil \frac{N-r}{\ell} \rceil)!} \ell^{\lceil \frac{N-r}{\ell} \rceil - 1} \right). \tag{6}$$

To get the matrix H_{gss} in standard form, it is actually enough to use the $\lceil \frac{N-r}{\ell} \rceil$ blocks for Π_1 so that the corresponding columns in H_{gss} contain r linearly independent columns.

Remark 3. We assumed that we have a parity check matrix H_{gss} of the form $[A|I]$. However, if this is not the case, we need to select the proper $\lceil \frac{N-r}{\ell} \rceil$ blocks containing r columns which are linearly independent. H_{gss} will then be transformed so that these columns will correspond to the identity matrix, and we can apply the same method as before.

5.2   Algebraic Cryptanalysis

Recall that, in the BGKR scheme, the μ-block structure is not preserved by the permutations, and the GRS code is public. Thus, the folding attack introduced by Faugere et al. [16] does not affect it.

We want to solve Eq. (3) algebraically. For that we also need to add more conditions on the entries of the permutation matrix Π. Let Π_1 and Π_2 be the two aforementioned permutation matrices such that $\Pi = \Pi_1 \Pi_2$. To properly model Π, we just need to provide a modelling of Π_1 and Π_2.

1. *Modelling of the matrix Π_1.* Let us denote by $(Y_{i,j})_{0 \leq i,j \leq s'-1}$ the entries of the matrix P_π. Then according to the definition of Π_1, its characterization is equivalent to that of P_π. Therefore equations describing properties of Π_1 as permutation matrix are just those of P_π. We have:

   • $\forall i \in [0, s' - 1]$, $\sum_{j=0}^{s'-1} Y_{i,j} = 1$

   • $\forall j \in [0, s' - 1]$, $\sum_{i=0}^{s'-1} Y_{i,j} = 1$

   • $\forall i \in [0, s' - 1]$, $\forall 0 \leq j < j' \leq s' - 1$, $Y_{i,j}Y_{i,j'} = 0$

   • $\forall j \in [0, s' - 1]$, $\forall 0 \leq i < i' \leq s' - 1$, $Y_{i,j}Y_{i',j} = 0$

2. *Modelling of the matrix Π_2.* For the purpose of our cryptanalysis we fix in the description of Π_2 the matrix $P_{shift}^{i_0}$ to be the identity matrix I_ℓ, that is $i_0 = 0$. Therefore, to characterize the permutation Π_2, we first need just to describe the matrices $P_{shift}^{i_\lambda}$ for all $\lambda = 1, 2, ..., s' - 1$.

   Note that the matrices $P_{shift}^{i_\lambda}$ are circulant for all $\lambda = 1, 2, ..., s' - 1$, and, as a consequence, we need just the first row of each of those matrices. Let us denote by $(X_{i,j}^{(\lambda)})_{\leq i,j \leq s'-1}$ the entries of $P_{shift}^{i_\lambda}$, for all $\lambda = 1, 2, ..., s' - 1$. Then, the equations describing the properties of Π_2 as permutation matrix are given by:

   • $\forall \lambda \in [1, s' - 1]$, $\sum_{j=0}^{\ell-1} X_{0,j}^{(\lambda)} = 1$

   • $\forall \lambda \in [1, s' - 1]$, $\forall 0 \leq j < j' \leq \ell - 1$, $X_{0,j}^{(\lambda)} X_{0,j'}^{(\lambda)} = 0$

By using the above modelling of Π_1 and Π_2 and the corresponding linear equations relative to the properties of the permutation matrix, we can reduce the number of unknowns in (3) and we are left with:

• $N_u = s'^2 + (s' - 1)\ell - 3s' + 2$ unknowns
• $N_e = (N - k') \times k_G + s'(s' - 1) + \frac{1}{2}\ell(\ell - 1)(s' - 1)$ quadratic equations,

where k_G is the number of vectors used to generate the public code quasi-cyclically. Our attack therefore consists in solving a system of quadratic equations in a certain number of variables. There are two approaches to solve this problem.

Using a Specialized Solver for Quadratic Boolean Equations. Since we only have to deal with Boolean equations, we consider the method proposed in [3] and we use it for the evaluation of the complexity in this paragraph. The method is closely related to Gröbner bases but it is completely specialized for quadratic Boolean systems. We can check that the number of equations $N_e = N^2(N - 1)$ is of the form $\alpha N'$, where $\alpha \geq 1$ and N' denotes the number of unknowns, and we have $\alpha > \frac{N^2(N-1)}{N^2} = N - 1$. Since we want a lower bound on the complexity, we take the exponent of linear algebra Θ (i.e. the cost of multiplying two $n \times n$ matrices is in $\mathcal{O}(n^\Theta)$) to be 2. We assume that the system is γ-strong regular for

$\gamma = 0.9 \cdot \alpha$ and we get that the complexity of solving our system is bounded by a function in $\mathcal{O}(2^{1-\gamma+\Theta \cdot F_\alpha(\gamma)})$. Thus, we have $\frac{\alpha}{\gamma} = \frac{1}{\epsilon}$, in a way that the parameter D of proposition 8 page 12 of [3] is given by $M(\frac{1}{\epsilon})$. The function M is defined by $M(x) = -x + \frac{1}{2} + \frac{1}{2}\sqrt{2x^2 - 10x - 1 + 2(x+2)\sqrt{x(x+2)}}$. If $D = M(\frac{1}{\epsilon})$, then $F_\alpha(\gamma) = -\gamma \log_2(D^D(1-D)^{1-D})$.

Using General Gröbner Basis. It is known that the complexity of solving such system of equations is given by

$$\mathcal{O}\left(\binom{N' + D_{reg}}{D_{reg}}^\Theta\right). \tag{7}$$

where D_{reg} is the degree of regularity of the system. There is no general method to estimate the degree of regularity of the Gröbner basis algorithm. However, in case the system is semi-regular, it satisfies $D_{reg} \leq 1 + N'$ (see for example [18]). Our system is not guaranteed to be semi-regular and, even if we assumed this was the case, the complexity of the algebraic attack would still be larger than the complexity of the exhaustive search attack. The main issue is that we have too many quadratic equations coming from the properties of the permutation matrices.

5.3 Parameter Choice

In Table 1, we give a summary of the complexity of the various KSA attacks on the parameters given in [12].

Table 1. Theoretical complexity of attacks on the parameters of [12] in the KSA model.

Parameters	Proposal 1	Proposal 2	Proposal 3	Proposal 4
m	12	12	13	14
ℓ	4095	4095	8191	5461
μ	9	10	11	11
s	1	1	1	2
t	897	897	1095	296
GRS Code	[4095,2300,1796]	[4095,2300,1796]	[8191,6000,2192]	[10922,5000,5923]
GSS Code	[36855,15315]	[40950,19410]	[90101,61618]	[120142,37234]
Public Code	[36855,4095]	[40950,4095]	[90101,8191]	[120142,5461]
Claimed	2^{128}	2^{128}	2^{128}	2^{196}
Ex. Search	2^{114}	2^{129}	2^{155}	2^{330}
Imp. Ex. Search	2^{47}	2^{62}	2^{113}	2^{104}
Non-KSA	2^{155}	2^{182}	2^{256}	2^{411}

In the last row, labeled "Non-KSA", we have presented the numbers relative to a more traditional attack, where we do not assume the conditions for the KSA model. We use the improved exhaustive search technique to guess the permutation matrices, and the projection matrix Ψ is obtained by taking random guesses.

Remark 4. It is natural to assess the security of the cryptosystem without the assumption of the KSA model. In (1), in addition to the permutation matrices, we are also looking for the projection matrices. So, in addition to the cost of the search for the permutation matrices, we have an additional factor in the complexity of the exhaustive search attack. The number of $m \times \mu$ binary matrices of rank μ is given by $N_{mat}(m,\mu) = \prod_{i=0}^{\mu-1}(2^m - 2^i)$ and hence this additional factor is given by $\left(\prod_{i=0}^{\mu-1}(2^m - 2^i)\right)^s$.

In total, performing an exhaustive search attack outside of the KSA model has the following complexity:

1. Exhaustive search: $\left(\prod_{i=0}^{\mu-1}(2^m - 2^i)\right)^s (s'!)\ell^{s'-1}$.
2. Improved exhaustive search: $\left(\prod_{i=0}^{\mu-1}(2^m - 2^i)\right)^s \dfrac{(s'!)}{(s' - \lceil\frac{N-r}{\ell}\rceil)!}\ell^{\lceil\frac{N-r}{\ell}\rceil - 1}$

This leads to a very high cost, and thus we can conclude that the original parameters are still secure without the KSA assumption. The security of the original scheme relies on the hidden projection matrices to get a subspace subcode.

As we can see in Table 1, the complexity of the improved exhaustive search is well below the claimed security bits, and therefore the original parameters are not secure in the KSA model. Thus, if one is concerned with such type of attacks, and want to guarantee security in this scenario, the parameters should be increased. We present a possible choice of such parameters in Table 2, where the aim is to reach 128 bits of security in the KSA model. For the example with $\ell = 5461$ in Table 2, in order to increase the security, we want to decrease the parameter r in Eq. 6. In order to do this, we need to increase the number of generators in the public code $\mathcal{C}_{pub}$. For the case where $\ell = 381$, we also do the same by using a larger $\mathcal{C}_{pub}$. It is immediate to note that such a choice causes a moderate increase in public-key size.

Table 2. New Parameters for 128 Security Bits in the KSA Model

m	ℓ	μ	s	t	GRS Code	GSS Code	Public Code	Key Size
14	381	9	17	1333	[6477,3810,2668]	[58293,20955]	[58293,4953]	6.6 Kbytes
14	5461	9	2	2730	[10922,5461,5462]	[98298,21844]	[98298,16383]	10.2 Kbytes

6   Conclusion

Code-based cryptography is one of the most accredited candidates for secure communication in the post-quantum era. Years of research building on the McEliece cryptosystem have led to several interesting variants; however, to date, the most promising direction for code-based encryption schemes that feature compact keys, is represented by systems based on structured parity-check codes such as QC-MDPC [1,5,30]. Compared to those, schemes based on "traditional" algebraic codes present some advantages – for example, the absence of a non-trivial Decoding Failure Rate (DFR) – but are not as competitive in term of data size. In this light, the work of BGKR is extremely interesting, as it offers the possibility of building a McEliece cryptosystem using algebraic codes, namely subspace subcodes of GRS codes, with an entirely innovative approach and achieving very small data sizes.

In this work, we have analyzed the security of the BGKR scheme to an additional degree, compared to the original work. We have introduced a new model, called Known Subspace Attack (KSA), which allows us to have a different perspective, and an in-depth look at the underlying security of the scheme. Thanks to our analysis we are able to show that under the KSA assumption, the original parameters were not secure, and we have thus suggested new parameters. Note that the new parameters cause a moderate increase in the size of the public key. We point out that, without the KSA assumption, the cryptosystem is not affected by the attacks described here, and the original parameters should still be considered secure.

References

1. Aragon, N., et al.: BIKE: bit flipping key encapsulation (2017). http://bikesuite.org/
2. Banegas, G., et al.: DAGS: key encapsulation using dyadic GS codes. J. Math. Cryptol. **12**(4), 221–239 (2018)
3. Bardet, M., Faugère, J., Salvy, B.: On the complexity of the F5 Gröbner basis algorithm. J. Symb. Comput. **70**, 49–70 (2015)
4. Barelli, É., Couvreur, A.: An efficient structural attack on NIST submission DAGS. In: Peyrin, T., Galbraith, S. (eds.) ASIACRYPT 2018. LNCS, vol. 11272, pp. 93–118. Springer, Cham (2018). https://doi.org/10.1007/978-3-030-03326-2_4
5. Barreto, P.S., et al.: Cake: code-based algorithm for key encapsulation. In: O'Neill, M. (ed.) Cryptography and Coding. LNCS, vol. 10655, pp. 207–226. Springer, Heidelberg (2017). https://doi.org/10.1007/978-3-319-71045-7_11
6. Barreto, P.S.L.M., Lindner, R., Misoczki, R.: Monoidic codes in cryptography. In: Yang, B.-Y. (ed.) PQCrypto 2011. LNCS, vol. 7071, pp. 179–199. Springer, Heidelberg (2011). https://doi.org/10.1007/978-3-642-25405-5_12
7. Berger, T.P., Cayrel, P.-L., Gaborit, P., Otmani, A.: Reducing key length of the McEliece cryptosystem. In: Preneel, B. (ed.) AFRICACRYPT 2009. LNCS, vol. 5580, pp. 77–97. Springer, Heidelberg (2009). https://doi.org/10.1007/978-3-642-02384-2_6
8. Berger, T.P., El Amrani, N.: Codes over $\mathcal{L}(\mathrm{GF}(2)^m, \mathrm{GF}(2)^m)$, MDS diffusion matrices and cryptographic applications. In: El Hajji, S., Nitaj, A., Carlet, C., Souidi, E. (eds.) Codes, Cryptology, and Information Security. LNCS, vol. 9084, pp. 197–214. Springer, Heidelberg (2015). https://doi.org/10.1007/978-3-319-18681-8_16
9. Berger, T.P., Gueye, C.T., Klamti, J.B.: Generalized subspace subcodes with application in cryptology. CoRR abs/1704.07882 (2017). http://arxiv.org/abs/1704.07882

10. Berger, T.P., Gueye, C.T., Klamti, J.B.: A NP-complete problem in coding theory with application to code based cryptography. In: El Hajji, S., Nitaj, A., Souidi, E.M. (eds.) C2SI 2017. LNCS, vol. 10194, pp. 230–237. Springer, Cham (2017). https://doi.org/10.1007/978-3-319-55589-8_15
11. Berger, T.P., Gueye, C.T., Klamti, J.B.: Generalized subspace subcodes with application in cryptology. IEEE Trans. Inf. Theory **65**(8), 4641–4657 (2019)
12. Berger, T.P., Gueye, C.T., Klamti, J.B., Ruatta, O.: Designing a public key cryptosystem based on quasi-cyclic subspace subcodes of Reed-Solomon codes. In: Gueye, C.T., Persichetti, E., Cayrel, P.-L., Buchmann, J. (eds.) A2C 2019. CCIS, vol. 1133, pp. 97–113. Springer, Cham (2019). https://doi.org/10.1007/978-3-030-36237-9_6
13. Berger, T.P., Loidreau, P.: How to mask the structure of codes for a cryptographic use. Designs Codes Crypt. **35**(1), 63–79 (2005). https://doi.org/10.1007/s10623-003-6151-2
14. Couvreur, A., Lequesne, M.: On the security of subspace subcodes of Reed-Solomon codes for public key encryption. arXiv preprint arXiv:2009.05826 (2020)
15. Drăgoi, V., Richmond, T., Bucerzan, D., Legay, A.: Survey on cryptanalysis of code-based cryptography: from theoretical to physical attacks. In: 2018 7th International Conference on Computers Communications and Control (ICCCC), pp. 215–223. IEEE (2018)
16. Faugère, J.-C., Otmani, A., Perret, L., de Portzamparc, F., Tillich, J.-P.: Structural cryptanalysis of McEliece schemes with compact keys. Designs Codes Crypt. **79**(1), 87–112 (2015). https://doi.org/10.1007/s10623-015-0036-z
17. Faugère, J.-C., Otmani, A., Perret, L., Tillich, J.-P.: Algebraic cryptanalysis of McEliece variants with compact keys. In: Gilbert, H. (ed.) EUROCRYPT 2010. LNCS, vol. 6110, pp. 279–298. Springer, Heidelberg (2010). https://doi.org/10.1007/978-3-642-13190-5_14
18. Faugère, J.C., Otmani, A., Perret, L., Tillich, J.P.: Algebraic cryptanalysis of compact McEliece's variants-toward a complexity analysis. In: Conference on Symbolic Computation and Cryptography, p. 45 (2013)
19. Gabidulin, E.M., Paramonov, A.V., Tretjakov, O.V.: Ideals over a non-commutative ring and their application in cryptology. In: Davies, D.W. (ed.) EUROCRYPT 1991. LNCS, vol. 547, pp. 482–489. Springer, Heidelberg (1991). https://doi.org/10.1007/3-540-46416-6_41
20. Gabidulin, E.M.: Attacks and counter-attacks on the GPT public key cryptosystem. Designs Codes Crypt. **48**(2), 171–177 (2008). https://doi.org/10.1007/s10623-007-9160-8
21. Gaborit, P., Murat, G., Ruatta, O., Zémor, G.: Low rank parity check codes and their application to cryptography. In: Proceedings of the Workshop on Coding and Cryptography WCC, vol. 2013 (2013)
22. Georgiades, J.: Some remarks on the security of the identification scheme based on permuted kernels. J. Cryptol. **5**(2), 133–137 (1992). https://doi.org/10.1007/BF00193565
23. Horlemann-Trautmann, A.L., Marshall, K., Rosenthal, J.: Extension of overbeck's attack for Gabidulin-based cryptosystems. Designs Codes Cryp. **86**(2), 319–340 (2018). https://doi.org/10.1007/s10623-017-0343-7
24. Huffman, W.C., Pless, V.: Fundamentals of Error-correcting Codes. Cambridge University Press, Cambridge (2010)
25. Khathuria, K., Joachim Rosenthal, V.W.: Encryption scheme based on expanded Reed-Solomon codes. Adv. Math. Commun. **15**(2), 207–218 (2021)

26. Loidreau, P.: Designing a rank metric based McEliece cryptosystem. In: Sendrier, N. (ed.) PQCrypto 2010. LNCS, vol. 6061, pp. 142–152. Springer, Heidelberg (2010). https://doi.org/10.1007/978-3-642-12929-2_11
27. McEliece, R., Van Tilborg, H.: On the inherent intractability of certain coding problems. IEEE Trans. Inf. Theory **24**(3), 384–386 (1978)
28. McEliece, R.J.: A public-key cryptosystem based on algebraic coding theory. Coding Thv **4244**, 114–116 (1978)
29. Misoczki, R., Barreto, P.S.L.M.: Compact McEliece keys from Goppa codes. In: Jacobson, M.J., Rijmen, V., Safavi-Naini, R. (eds.) SAC 2009. LNCS, vol. 5867, pp. 376–392. Springer, Heidelberg (2009). https://doi.org/10.1007/978-3-642-05445-7_24
30. Misoczki, R., Tillich, J.P., Sendrier, N., Barreto, P.S.: MDPC-McEliece: new McEliece variants from moderate density parity-check codes. In: 2013 IEEE International Symposium on Information Theory, pp. 2069–2073. IEEE (2013)
31. Niederreiter, H.: Knapsack-type cryptosystems and algebraic coding theory. Prob. Control Inf. Theory **15**(2), 159–166 (1986)
32. NIST. https://csrc.nist.gov/projects/post-quantum-cryptography/round-3-submissions
33. NIST: https://csrc.nist.gov/Projects/post-quantum-cryptography/round-2-submissions
34. Overbeck, R.: Structural attacks for public key cryptosystems based on Gabidulin codes. J. Cryptol. **21**(2), 280–301 (2008). https://doi.org/10.1007/s00145-007-9003-9
35. Persichetti, E.: Compact McEliece keys based on quasi-dyadic Srivastava codes. J. Math. Cryptol. **6**(2), 149–169 (2012)
36. Rashwan, H., Gabidulin, E.M., Honary, B.: A smart approach for GPT cryptosystem based on rank codes. In: 2010 IEEE International Symposium on Information Theory, pp. 2463–2467. IEEE (2010)
37. Shor, P.W.: Algorithms for quantum computation: discrete logarithms and factoring. In: Proceedings 35th Annual Symposium on Foundations of Computer Science, pp. 124–134. IEEE (1994)
38. Sidelnikov, V.M., Shestakov, S.O.: On insecurity of cryptosystems based on generalized Reed-Solomon codes (1992)
39. Wieschebrink, C.: Cryptanalysis of the Niederreiter public key scheme based on GRS subcodes. In: Sendrier, N. (ed.) PQCrypto 2010. LNCS, vol. 6061, pp. 61–72. Springer, Heidelberg (2010). https://doi.org/10.1007/978-3-642-12929-2_5

Information-Set Decoding with Hints

Anna-Lena Horlemann[1]([✉]), Sven Puchinger[2], Julian Renner[2],
Thomas Schamberger[2], and Antonia Wachter-Zeh[2]

[1] School of Computer Science, University of St. Gallen, St. Gallen, Switzerland
anna-lena.horlemann@unisg.ch
[2] Technical University of Munich (TUM), Munich, Germany
{sven.puchinger,julian.renner,t.schamberger,antonia.wachter-zeh}@tum.de

Abstract. This paper studies how to incorporate small information leakages (called "hints") into *information-set decoding* (ISD) algorithms. In particular, the influence of these hints on solving the (n, k, t)-*syndrome-decoding problem* (SDP), i.e., generic syndrome decoding of a code of length n, dimension k, and an error of weight t, is analyzed. We motivate all hints by leakages obtainable through realistic side-channel attacks on code-based post-quantum cryptosystems. One class of studied hints consists of partial knowledge of the error or message, which allow to reduce the length, dimension, or error weight using a suitable transformation of the problem. As a second class of hints, we assume that the Hamming weights of sub-blocks of the error are known, which can be motivated by a template attack. We present adapted ISD algorithms for this type of leakage. For each third-round code-based NIST submission (Classic McEliece, BIKE, HQC), we show how many hints of each type are needed to reduce the work factor below the claimed security level. E.g., for Classic McEliece `mceliece348864`, the work factor is reduced below 2^{128} for 9 known error locations, 650 known error-free positions or known Hamming weights of 29 sub-blocks of roughly equal size.

Keywords: Post-quantum cryptography · Code-based cryptography · Information set decoding · Side-channel attacks

1   Introduction

Shortly after the proposal of the first public-key cryptosystem [14], Berlekamp *et al.* proved that decisional decoding in a random linear code is an NP-complete problem [7]. In the same year, McEliece designed the first encryption scheme that

This work was supported by the German Research Foundation (Deutsche Forschungsgemeinschaft, DFG) under Grant No. WA3907/4-1 and SE2989/1-1 and the European Research Council (ERC) under the European Union's Horizon 2020 research and innovation programme (grant agreement No 801434). This work was partly done while S. Puchinger was with the Technical University of Denmark, Lyngby, Denmark, where he was supported by the European Union's Horizon 2020 research and innovation programme under the Marie Skłodowska-Curie grant agreement no. 713683.

© Springer Nature Switzerland AG 2022
A. Wachter-Zeh et al. (Eds.): CBCrypto 2021, LNCS 13150, pp. 60–83, 2022.
https://doi.org/10.1007/978-3-030-98365-9_4

relies on the difficulty of the aforementioned problem. However, due to practical issues of the McEliece cryptosystem (i.e., relatively large key sizes), other systems were usually employed, e.g., schemes based on the hardness of factoring large integers and computing discrete logarithms. However, Shor's quantum algorithm [33] will be capable of factoring large integers and computing discrete logarithms efficiently as soon as a capable quantum computer exists. Code-based cryptography is believed to resist attacks by quantum computers and thus, the Niederreiter-variant of the McEliece scheme is one of the four finalists in the third round of the *National Institute of Standards and Technology* (NIST) [25] post-quantum security standardization process. In addition, there are two alternate code-based candidates in the third round, namely BIKE and HQC.

Since the schemes in the third round seem to be secure from a theoretical point of view, more and more investigations with respect to their side-channel secure implementation are conducted. With a so called side-channel attack, an attacker is able to exploit additional information obtained through observing the execution of a cryptographic algorithm in order to retrieve secret information, e.g., the private key. The two most prominent variants are timing attacks, where an attacker exploits execution time differences dependent on the secret, and power attacks, where the attacker measures the power consumption of the executing device as it is data dependent for CMOS logic. For the original McEliece paper [22], there are several attacks [10,23,34,37,38] including attacks against implementations that use a non-constant time Patterson decoder. For the submission Classic McEliece, a reaction attack utilizing an electromagnetic (EM) side-channel that leads to a full message recovery is shown in [19]. The submission BIKE is based on a quasi-cyclic code that allows for an optimized multiplication routine for the matrix vector multiplication during the syndrome computation. Published attacks [31,35,39] target this multiplication, where [39] successfully attacks a non-constant-time variant and [31,35] target a constant-time version of the multiplication proposed in [11]. The third code-based cryptosystem HQC has been attacked in [27,40], where the authors exploit the non-constant time implementation of the used BCH decoder. In [32] a successful power side-channel attack against a constant-time decoder (proposed as a countermeasure in [40]) is shown.

Most of these attacks are able to directly retrieve the entire secret key or plaintext. In [31], a tailored approach for specific partial leakage is given. In this paper, we provide a more general approach and show how arbitrary small information leakage or partial attack results can be incorporated into algorithms that solve the general decoding problem for linear codes and how this affects the security level of code-based cryptosystems. For that, we adapt the theoretical ideas of [13] from lattice-based systems to the code-based setting. More precisely, the authors of [13] propose a framework that can handle four different types of side information (called *hints*) to increase the efficiency of lattice reduction techniques. We propose a similar framework for code-based cryptography, where we translate concepts of these hints to the code-based scenario, introduce additional hints, and show how to transform the decoding problem accordingly.

Some previous works consider partial leakage. In [19,32] it was shown how partial information from attacks can be used to reduce the complexity of the underlying hard mathematical problem (namely, the *syndrome decoding problem* (SDP)). The reason for obtaining only partial information in these attacks is either a limitation on the obtainable side-channel observations as in [19] or certain private keys that can only partially be retrieved with the proposed attack as in [32]. The knowledge of some plaintext bits or some error bits and its impact on the performance of *information set decoding* (ISD) was already considered, even if only briefly, in [10]. In [34] and [19], side-channel measurements (or timing attacks) are used to determine (some of) the error positions. In the former, all error positions are found by measurements, whereas in the latter, partial knowledge of these positions is combined with an ISD algorithm. Furthermore, the latter algorithm is generalized to work with the knowledge of a set of indices containing some errors, or vice versa, the knowledge of error-free positions. In [26] two hints are considered: the error values come from a subset or the entries of the error vector are known, but their positions are not. They adapt an ISD algorithm to incorporate these hints. It is left as an open question how to incorporate these hints into generalized birthday decoding algorithms. The cryptosystem set up in [15] uses error vectors whose sub-blocks are from a known set of short vectors. The corresponding generator matrices of the used error-correcting codes have a prescribed triangular block structure. Both these facts are used as partial knowledge in the ISD algorithm for cryptanalyzing this system in [15] and [24]. In [5] a zero-knowledge identification scheme was set up, using errors which live in a subset of the generally used finite field. For analyzing the strength of this scheme the authors set up several ISD algorithms exploiting the fact that the error entries were restricted.

In this paper, we provide a general framework to incorporate partial information leakages (hints) into ISD algorithms. The hints that we consider include general considerations like knowing parts of the message or a measurement of the message, some erroneous or error-free locations, or knowing the Hamming weight of blocks of the error. For each hint, we provide a motivation from the side-channel perspective in the respective chapter. As a general motivation, partial knowledge of the error or message allows an attacker to cope with restrictions on the maximum amount of side-channel observations or to simplify the attack by only allowing the retrieval of partial results in certain special cases. The hint of knowing the Hamming weight for certain error blocks is motivated by practical template attacks [12]. To reduce the amount of required side-channel observations, an attacker might choose to use Hamming weight templates instead of value templates. This might even be a necessity if more than 16-bit value templates have to be created for a successful attack. For each hint, we show how the SDP is transformed into an SDP with smaller parameters, which therefore can be solved with smaller complexity. We apply these hints on Classic McEliece, HQC and BIKE and show how much leakage of each type is required to reduce the logarithmic work factor below the claimed security level.

The paper is organized as follows. In the following section we define some notation, state the syndrome decoding problem (SDP), and give the parameters of the SDP instances from the three code-based submissions to the third round of the NIST standardization project that we consider in this paper. Moreover, we give a very brief overview of the classical information set decoding (ISD) algorithms we will use and adapt to our settings. In Sect. 3 we show how simple hints, like partial knowledge of the error or the message, reduce the parameters of (and hence simplify) a given SDP. Using Stern's decoding algorithm, we show in Fig. 1 how these reductions effect the security of the NIST submissions Classic McEliece and BIKE. In Sect. 4 we consider the setting where we know the exact weight decomposition of the error vector into blocks of (approximately) equal length. We adapt known ISD algorithms to incorporate this extra knowledge and show how this effects the security level of the NIST submissions, depending on the number of blocks for the weight decomposition of the error vector. We conclude this paper in Sect. 5.

2 Preliminaries

Let $[n] := \{1, 2, \ldots, n\}$ and let $\mathbb{F}_q$ denote the finite field of order q. Matrices and vectors are denoted by A and a, respectively, and their entries are denoted by A_{ij} and a_i, respectively, where $i, j \geq 1$. To avoid confusion with the error vectors, δ_i denotes the i-th unit vector. For a given set of indices I and a matrix A, the matrix A_I denotes the submatrix of A containing the rows indexed by I. For a vector x, x_I denotes the subvector containing the entries indexed by I.

The Hamming weight wt_H of a vector $x = [x_1, \ldots, x_n] \in \mathbb{F}_q^n$ is defined by $\mathrm{wt}_H(x) = |\{x_i \neq 0 \mid i \in [n]\}|$. The parameters of a linear code $\mathcal{C} \subseteq \mathbb{F}_q^n$ of length n, dimension k and minimum distance d are denoted by $[n, k, d]_q$. Let $G \in \mathbb{F}_q^{k \times n}$ and $H \subset \mathbb{F}_q^{(n-k) \times n}$ be a generator matrix and a parity-check matrix, respectively, of an $[n, k, d]_q$-code $\mathcal{C}$. If $m \in \mathbb{F}_q^k$ denotes a message and $e \in \mathbb{F}_q^n$ an error vector, then $c = mG \in \mathcal{C}$ is the corresponding codeword and $r = mG + e$ is the corresponding received word. The corresponding syndrome $s \in \mathbb{F}_q^{n-k}$ with respect to H is given by $s = rH^\top = eH^\top$. The goal of syndrome decoding is to find the lowest-weight error vector e that fulfills the above syndrome equation. The generic (and decisional) version of this *syndrome decoding problem* is known to be NP-complete [7] and is the basis of most code-based cryptosystem. The computational version with prescribed weight t can be formulated as:

Definition 1. $((n, k, t)$-Syndrome Decoding Problem (SDP)). *Given $H \in \mathbb{F}_q^{(n-k) \times n}$, $t \in \mathbb{N}$ and $s \in \mathbb{F}_q^{n-k}$, find $e \in \mathbb{F}_q^n$ such that $\mathrm{wt}_H(e) = t$ and $eH^\top = s$.*

With the notation above, we can state the problem also in generator matrix form: Given G and r, find e of weight t such that $r - e \in \mathcal{C}$. We can switch between generator and parity-check matrix via linear algebra, and can compute s from r, and vice versa if we are only interested in e and not a specific codeword $r - e$. We need to be careful about the latter restriction in Sect. 3, where we assume that we have partial knowledge about $r - e$.

Throughout this paper, we exemplify our results with different parameter sets of all code-based submissions to the third round of the NIST standardization: Classic McEliece, BIKE, and HQC [1–3]. The security of Classic McEliece and BIKE can be reduced to an (n, k, t)-SDP through a message attack.

For HQC, additional information can directly be included into generic decoders by system design. Therefore, one can perform a key attack based on any algorithm that solves the (n, k, t)-SDP with additional information: The resulting solution of the (n, k, t)-SDP should have Hamming weights exactly $t/2$ in e_1 and e_2, respectively, where $e_1, e_2 \in \mathbb{F}_2^{n/2}$ are two equally-sized blocks of the error $e = [e_1, e_2]$. Therefore, the additional information is the Hamming weight of the two blocks. This is true for HQC by design. Additional knowledge from side-channel attacks can further improve that principle.

The parameters (n, k, t) of the considered instances of the SDP are as follows.

Parameter Set	n	k	t	Security Level	Used for
mceliece348864	3488	2720	64	NIST Cat. 1 (128)	Message attack
mceliece6688128	6688	5024	128	NIST Cat. 5 (256)	Message attack
BIKE-Level-1	24646	12323	134	NIST Cat. 1 (128)	Message attack
BIKE-Level-5	81946	40973	264	NIST Cat. 5 (256)	Message attack
hqc-128	35338	17669	132	NIST Cat. 1 (128)	Key attack
hqc-256	115274	57637	262	NIST Cat. 5 (256)	Key attack

The table also contains the claimed security levels. Note that the parameters were designed with a security margin. E.g., Stern's information-set decoder (see below) has a work factor of $\approx 2^{146}$ for mceliece348864, which is ca. 18 bits above the claimed security level.

The idea of *information-set decoding* (ISD) was first introduced by Prange [29]. In this algorithm, in each iteration one chooses a random information set, reconstructs a message from the corresponding entries, then re-encodes this message and checks if the resulting codeword has Hamming distance at most t to the received word. The average computational complexity is roughly $\binom{n}{t}/\binom{n-k}{t}(n-k)^2(n+1)$ binary operations. A more advanced ISD variant is the one by Lee and Brickell [20], where some errors in the information set are possible. Even more advanced is Stern's ISD algorithm [36]. It partitions the errors in the information set into two parts of prescribed weight each, and moreover fixes a zero window in the remaining error vector. We will use Stern's algorithm, sped up with intermediate sums and early abort techniques, to attack the various cryptosystems with the hints described in the following sections. We remark that other generic algorithms like MMT [21] or BJMM [6] might be slightly faster than the above. We describe a general framework for such birthday decoders in the appendix. The exact analysis and optimization of free parameters in these algorithms is out of the scope of this paper, and will be left as an open question for future work.

3 Hint-Based Parameter Reduction in the SDP

We introduce a class of hints and present, for all hints, generic transformations that reduce the SDP parameters. Hence, any ISD algorithm can be applied to the transformed instance to reduce the work factor. We exemplify the impact of these hints on the NIST proposals in Fig. 1 below and determine the number of hints needed to reduce the security levels below the claimed level. In Subsect. 3.6, we show how different hints can be combined in any order, analogous to the hints considered in the lattice setting in [13].

3.1 Known Error Locations or Error-Free Locations

First, we treat the setting where we gain some knowledge about the error vector: (i) either we know that a given position contains an error (and here we distinguish again if we also know the error value or not), or (ii) that the position is error-free. The case of knowing an error location in the binary case (which implies knowledge about the error value) and its implication on Classic McEliece was already studied in [19]; however, for comparison we include it in this section. These types of hints can be obtained from attacks that directly target the error. In [19], an attack on Classic McEliece is shown for which an electromagnetic (EM) side-channel oracle is constructed in order to distinguish if decoding is successful. Using this oracle the authors successively add an error to each position of the ciphertext and observe the oracle for each decryption. If the position was error-free, an additional error is induced, which results in an error during decoding indicated by the oracle. The authors provide a trade-off between required attack measurements and remaining attack complexity by utilizing the partial information (only a part of the error locations is known) in an ISD algorithm. A similar attack against HQC is shown in [32][1], where the authors retrieve the private key by observing the decoding result in the power side-channel for specially crafted ciphertexts. In the case of a successful attack, the exact error positions of the secret key are known, but the overall attack success is dependent on the support of the error. The authors therefore provide an attack that is able to retrieve the exact error positions of only a large part of possible keys. For the remaining keys, the error positions can only be partially retrieved. Nevertheless, these error positions can again be interpreted as hints.

Hint Type 1. *Consider an (n, k, t)-SDP with given parity-check matrix $\boldsymbol{H}$ and syndrome $\boldsymbol{s}$, which has a solution $\boldsymbol{e}$ such that $\mathrm{wt}_H(\boldsymbol{e}) = t$ and $\boldsymbol{e}\boldsymbol{H}^\top = \boldsymbol{s}$.*
Given: *An error entry $e_j \in \mathbb{F}_q \backslash \{0\}$.*

---

[1] The attack in [32] works for the original construction used in the HQC submission (a concatenation of a BCH code and a repetition code), but still has to be adapted to the latest HQC setting which uses a concatenation of a Reed-Muller and Reed-Solomon code, as the authors of HQC shifted to this code combination in their third round submission.

Theorem 1. *Using Hint Type 1, any (n, k, t)-SDP can be transformed into an $(n-1, k-1, t-1)$-SDP or an $(n-1, k, t-1)$-SDP. For a random code the former happens with high probability.*

Proof. If we know the error value e_j, then we can shorten the code in the j-th position and solve the new syndrome equation $e_{[n]\setminus\{j\}}(H_{[n]\setminus\{j\}})^\top = s - e_j H_j^\top$, where $e_{[n]\setminus\{j\}}$ has Hamming weight $t - 1$. For a random code and large n, the corresponding shortened code has dimension $k - 1$ with high probability, and we hence have a $(n - 1, k - 1, t - 1)$-SDP; otherwise the shortened code has dimension k, and we hence have a $(n - 1, k, t - 1)$-SDP. $\square$

Hint Type 2. *Consider an (n, k, t)-SDP with given parity-check matrix H and received word r, which has a solution e such that $\mathrm{wt}_H(e) = t$ and $eH^\top = rH^\top$.*
Given: An error location j with $e_j \neq 0$.

Theorem 2. *Using Hint Type 2, any (n, k, t)-SDP can be transformed into an $(n-1, k, t-1)$-SDP or an $(n-1, k-1, t-1)$-SDP. For a random code the former happens with high probability.*

Proof. If we know an error location j, but not the corresponding error value, then we can puncture the code in the j-th position, compute the corresponding parity-check matrix $H' \in \mathbb{F}_q^{(n-k-1)\times(n-1)}$ and solve the new syndrome equation $e_{[n]\setminus\{j\}}(H')^\top = r_{[n]\setminus\{j\}}(H')^\top$, where $e_{[n]\setminus\{j\}}$ has Hamming weight $t - 1$. As above, this leads to an $(n - 1, k, t - 1)$-SDP or an $(n - 1, k - 1, t - 1)$-SDP. Then, the value e_j can be found by simple erasure decoding, e.g., by Gaussian elimination. $\square$

If we know several, say ϵ many, error locations, indexed by the set $I \subset [n]$, we can extend the above ideas by shortening or puncturing the code in I.

Corollary 1. *With 9 known error locations (i.e., about 14%) the security level of* mceliece348864 *reduces to less than $126.38 < 128$ bits. For* mceliece6688128 *the security level reduces to less than $254.57 < 256$ bits with 4 known error locations (i.e., about 3%).*

We will now consider knowledge about error-free positions.

Hint Type 3. *Consider an (n, k, t)-SDP with given parity-check matrix H and syndrome s, which has a solution e such that $\mathrm{wt}_H(e) = t$ and $eH^\top = s$.*
Given: An error-free location, i.e., an index j such that $e_j = 0$.

Theorem 3. *Using Hint Type 3, any (n, k, t)-SDP can be transformed into an $(n - 1, k - 1, t)$-SDP or an $(n - 1, k, t)$-SDP. For a random code the former happens with high probability.*

Proof. The proof is analogous to the one of Theorem 1, with $e_j = 0$. $\square$

Corollary 2. *With 652 known error-free locations (i.e., about 19%) the security level of* mceliece348864 *reduces to less than $127.98 < 128$ bits security, using the reduction from Theorem 3. For* mceliece6688128 *the security level reduces to less than $255.99 < 256$ bits with 249 known error-free locations (i.e., about 4%).*

3.2   Measurement of the Error

During the encryption process of many code-based cryptosystems, one needs to compute the syndrome s of an error e w.r.t. the public parity-check matrix H, where all vectors/matrices are in a finite field, typically $\mathbb{F}_2$. This is done using a vector-matrix multiplication $s = eH^{\top}$. In practice, this multiplication can be done by grouping entries of the vectors into blocks of a fixed bit size B (e.g., $B = 8$ bits in the reference implementation of Classic McEliece [2]). This means that for each row H_i of H and each $j = 0, \ldots, B - 1$, we compute in parallel the j-th shifted inner product

$$b_{i,j} := \sum_{\mu=0}^{n/B-1} e_{j+\mu B} H_{i,j+\mu B}.$$

To get the inner product of e with the i-th row of H, we finally need to sum up

$$eH_i^{\top} = \sum_{j=0}^{B-1} b_{i,j}.$$

If we are able to retrieve, for some i, the B-bit vector $b_i = |b_{i,0}, \ldots, b_{i,B-1}|$, then we obtain B measurements of the error, where the measurement vectors are of the form

$$v_{i,0} := [H_{i,0}, \underbrace{0, \ldots, 0}_{B-1 \text{ zeros}}, H_{i,B}, 0, \ldots, 0, H_{i,2B}, 0, \ldots, 0, H_{i,n-B}]$$

$$v_{i,1} := [0, H_{i,1}, 0, \ldots, 0, H_{i,B+1}, 0, \ldots, 0, H_{i,2B+1}, 0, \ldots, 0]$$

$$\vdots$$

$$v_{i,B-1} := [0, \ldots, H_{i,B-1}, 0, \ldots, 0, H_{i,2B-1}, 0, \ldots, 0, H_{i,n-1}].$$

The vector b_i, for some i, can be obtained through a template attack on the syndrome computation in the encryption step. Please note that this attack on the reference implementation of Classic McEliece allows for the usage of value templates since B is small.

Any linear operation on the error vector entries that can be observed exactly (i.e., the exact result of the operation can be obtained), can in principle be used for this hint. For instance, if we can obtain intermediate results of a chunk-based syndrome computation in Niederreiter variants of the McEliece cryptosystem (as in the reference implementation of Classic McEliece [2]), we obtain measurements of the error directly.

Hint Type 4. *Consider an (n, k, t)-SDP with given parity-check matrix H and syndrome s, which has a solution e such that $\mathrm{wt}_H(e) = t$ and $eH^{\top} = s$.*
Given: A measurement vector $v \in \mathbb{F}_q^n \setminus \{0\}$, s.t. $v \notin \mathcal{C}^{\perp}$, and $e \cdot v^{\top} = \sigma \in \mathbb{F}_q$.

Recall that the kernel of H is the code $\mathcal{C}$. Hence, the condition $v \notin \mathcal{C}^{\perp}$ is equivalent to the condition that v and the rows of H are linearly independent.

Theorem 4. *Using Hint Type 4, any (n, k, t)-SDP can be transformed into an $(n, k - 1, t)$-SDP.*

Proof. Assume a *measurement vector* $v \in \mathbb{F}_q^n$ is given such that $v \notin \mathcal{C}^\perp$ and $e \cdot v^\top = \sigma$, then we can extend the original syndrome equation to $\bar{s} := [s \ \sigma] = e[H^\top \ v] =: e\bar{H}^\top$. Hence, we obtain a new parity-check matrix $\bar{H}^\top \in \mathbb{F}_q^{(n-k+1)\times n}$ and a corresponding syndrome $\bar{s} \in \mathbb{F}_q^{n-k+1}$. Since $v \notin \mathcal{C}^\perp$, we have $\mathrm{rank}(\bar{H}) = n - k + 1$, so $\bar{H}$ is a parity-check matrix of a subcode of $\mathcal{C}$ of dimension $k - 1$. □

When combining several measurements for the error, one needs to check that all of them, together with the rows of H, are linearly independent. Otherwise, not all of them will reduce the SDP to a smaller instance.

Corollary 3. *With 175 known linearly independent error measurements, the security level of mceliece348864 reduces to less than $127.97 < 128$ bits security. For mceliece6688128 the security level reduces to less than $255.96 < 256$ bits with 64 linearly independent error measurements.*

A special instance of an error measurement is the knowledge of an entry of the error vector, by setting v as the unit vector with the only non-zero entry in the error location, and σ as the error value. It depends on the code parameters and the generic decoder used, which reduction is more useful, see also Fig. 1. In particular, when applied to known error-free locations in *Classic McEliece*, Theorem 4 provides a stronger reduction of the security level than the reduction from Theorem 3.

3.3   Known Partial Message

This hint is only applicable for the SDP given in its generator matrix form, where we consider G instead of H and r instead of s as, e.g., in the original paper of the McEliece cryptosystem [22], where the ciphertext is of the form $r = mG + e$. In this system an attacker may obtain information about the message m by observing the encryption. A possible attack vector is a template attack on the load operations of the different parts of the message. E.g., the correct retrieval of an 8-bit part of the message gives a total of eight hints.

Hint Type 5. *Consider an (n, k, t)-SDP in generator matrix form (given G and $r = mG + e$), which has a solution e with $\mathrm{wt}_H(e) = t$.*
Given: A message entry $m_j \in \mathbb{F}_q$.

Theorem 5. *Using Hint Type 5, any (n, k, t)-SDP can be transformed into an $(n, k - 1, t)$-SDP.*

Proof. If an entry m_j of the message m is known, we can reduce the (n, k, t)-SDP to an $(n, k - 1, t)$-SDP. This reduction can be done by taking the subcode corresponding to the unknown message bits, computing the corresponding parity-check matrix of size $(n - k + 1) \times n$ and syndrome-decoding the modified received word $r' := r - m_j G_j$ with respect to the $(n - k + 1) \times n$ parity-check matrix. □

Note that, although coming from different types of hints, the reductions in Theorem 4 and Theorem 5 are mathematically the same.

Corollary 4. *With* 175 *known message entries (about* 6.5%) *the security level of the original McEliece system [22] in generator matrix form with the parameters proposed in* mceliece348864 *reduces to less than* 127.97 < 128 *bits. For the parameters suggested in* mceliece6688128 *the security level reduces to less than* 255.96 < 256 *bits with* 64 *known message entries (about* 1%).

3.4   Measurement of the Message

The motivation of this hint works analogously to the measurement of the error described in Subsect. 3.2. This means that the knowledge of every exact result of a linear operation on the message vector entries can be used for this hint. If the matrix vector operation mG is again implemented as a chunk-based multiplication, a successful template attack on the final multiplication result of one chunk can be used as this hint.

Hint Type 6. *Consider an* (n, k, t)-*SDP in generator matrix form (given* G *and* $r = mG + e$), *which has a solution* e *with* $\mathrm{wt}_{\mathrm{H}}(e) = t$. *Given: A measurement vector* $v \in \mathbb{F}_q^k \backslash \{0\}$ *such that* $m \cdot v^\top = \sigma \in \mathbb{F}_q$.

Theorem 6. *Using Hint Type 6, any* (n, k, t)-*SDP can be transformed into an* $(n, k - 1, t)$-*SDP.*

Proof. If we have $m \cdot v^\top = \sigma$, we can use the fact that there is some invertible $A \in \mathbb{F}_q^{k \times k}$ such that $Av^\top = \delta_k$, which gives $mA^{-1} \cdot Av^\top = \bar{m} \cdot \delta_k^\top = \bar{m}_k = \sigma$, where $\bar{m} := mA^{-1}$. Then, we get $mG = \bar{m}(AG)$ and $\bar{m}_k = \sigma$. We only need to find a solution $\bar{m}_{[k-1]}$ to the SDP given by the generator matrix $\bar{G} := (AG)_{[k-1]}$ and the transformed received word $\bar{r} := r - \sigma(AG)_k$. This gives an $(n, k - 1, t)$-SDP. □

Note that the reduction from Theorem 6 is mathematically again the same as the one in Theorem 4 and Theorem 5. When several measurement vectors are given that span an ϵ-dimensional subspace, the (n, k, t)-SDP can be reduced to an $(n, k - \epsilon, t)$-SDP, with the same techniques as above.

Corollary 5. *With* 175 *linearly independent message measurements, the security level of* mceliece348864 *reduces to less than* 127.97 < 128 *bits security. For* mceliece6688128, *the security level reduces to less than* 255.96 < 256 *bits with* 64 *linearly independent message measurements.*

Note that measurement of the message vector is a generalization of knowing a message symbol. Here, the two hints lead to the same reduction, in contrast to the analog measurement/known entry of the error.

Figure 1 exemplifies the work factor reduction of Stern's algorithm (work factor formula as in [17]) for all discussed hints, for Classic McEliece and BIKE.

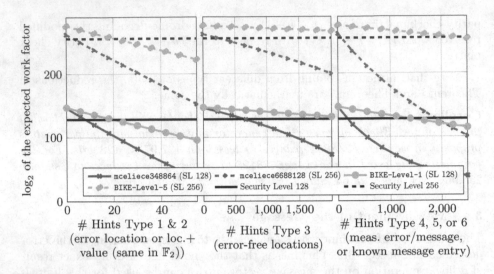

Fig. 1. Influence of hints on work factor of Stern's ISD algorithm, exemplified for two parameter sets each of the NIST round 3 submissions Classic McEliece and BIKE.

3.5 Restricted Error Values

Consider a cryptosystem over $\mathbb{F}_q$ with $q > 2$ which is designed such that the error values are from a proper subset $E \subset \mathbb{F}_q$. This knowledge can again speed up any decoder. The randomly guessed (partial) error vectors in the decoding algorithm are now chosen from $E^{n'}$ instead of $\mathbb{F}_q^{n'}$ (where n' is the length of the respective partial error vector). This has, e.g., been done in [5], where the error vectors are in $\{\pm 1\}^n$.

This idea can straightforwardly be extended if one only knows that a part of the error vector has restricted error values.

Remark 1. Since Classic McEliece, BIKE and HQC are defined over $\mathbb{F}_2$, this consideration has no impact on their security.

3.6 Combining Different Types of Hints

The restricted error values from Subsect. 3.5 can easily be combined with any other hint described previously, since they do not transform the parameters of the SDP, and are only used within the algorithms solving any given SDP. Therefore, we will now focus on how to combine the results from Subsects. 3.1 to 3.4.

If we know an error measurement (as in Subsect. 3.2) or a message measurement (as in Subsect. 3.4), which includes the case of knowing a message entry (as in Subsect. 3.3), we can transform the original SDP into one where the sought-after codeword lives in a subcode of the original code, with the same error vector, but different received word. Since the error vector is unchanged in this new SDP instance, any hint about it can be used in the usual way. Note however, that when combining message measurements and/or error measurements, it might happen that not all of the hints are linearly independent and that thus some of them do not add any extra information.

If we know a precise error value (as in the first and third case in Subsect. 3.1), say e_j, then we shorten the code and delete the j-th coordinate in the error vector and the received word (and if $e_j \neq 0$ we decrease the weight of the error vector). The shortening implicitly contains the hint $r_j - e_j = mG(j)$, where $G(j)$ denotes the j-th column of G, which is again a measurement of the message. It can hence be combined with other measure measurements (or error measurements) as above, in the shorter code.

If we know only an error location but not the value (as in the second case in Subsect. 3.1), then we puncture the original code and do not have the implicit measurement which reduces the dimension of the code. That means that m is not changed in the reduction, and hence any hints about the message can be used, also in the reduced $(n-1, k, t-1)$-SDP.

Overall, it follows that all hints can be combined with each other, in arbitrary order. However, it might be that not all of them are linearly independent to prior hints, and might not provide any extra information, and hence no reduction of the SDP.

4 Hamming Weights of Error Blocks Known

We mathematically model this type of leakage as follows: Partition the set $\mathcal{W} = \{1, \ldots, n\}$ into subsets (called blocks) $\mathcal{W}_i$ of cardinality η_i for $i = 1, \ldots, \ell$, and write $\boldsymbol{\eta} := [\eta_1, \ldots, \eta_\ell]$. We assume in the following that we know the exact weight decomposition t of the errors into the sets $\mathcal{W}_i$, i.e., the Hamming weights $t_i := \mathrm{wt}_H(e_{\mathcal{W}_i})$ of the errors restricted to the sets $\mathcal{W}_i$, for all $i = 1, \ldots, \ell$. These types of hints are motivated by template attacks on operations containing the error, e.g., load and store operations of continuous bits (blocks) of the error vector. If the size of simultaneously processed error bits becomes too large (a conservative estimate would be larger than 16 bit), the profiling step for all possibilities of values is not practical anymore. In this case, and also to reduce the attack complexity for smaller template sizes, an attacker can opt to use only Hamming weight templates. In this case, a successful template matching reveals the Hamming weight of the targeted block. An example of such an attack can be given with the HQC cryptosystem. The first step during the HQC decryption consists of the subtraction $v - u\mathrm{rot}(y)^\top$ (see [1, page 9]), where v and u define the ciphertext and y is a sparse error vector which constitutes the private key. By fixing $v = [0 \ldots 0]$ and $u = [1\ 0 \ldots 0]$ the private key is subtracted from a zero vector. A template attack using Hamming weight templates is then able to retrieve the Hamming weight of each subtraction block. The block size depends on the register size of the target platform.

In the following, we adapt well-known ISD algorithms to include this information leakage and describe how this reduces the security level. We remark that ISD algorithms for prescribed block weights were also studied in [4,9], however only for regular error vectors, i.e., where the blocks of the error vector have the same length and the same weight. Those algorithms are hence not applicable to our setting, where we assume that the error vector is uniformly chosen among all vectors of weight t, and we are able to gain some knowledge about the distribution of the errors.

In this paper, we focus on Prange's, Lee and Brickell's, and Stern's algorithms. An adaption to Wagner's algorithm [41] is outlined in Appendix A. Further, we consider all algorithms over $\mathbb{F}_2$ since this is the most relevant case for cryptography and the three considered NIST submissions are all over $\mathbb{F}_2$. A generalization to arbitrary fields $\mathbb{F}_q$ is possible similar to [18,28]. Note that for the considered algorithms, the expected number of iterations is independent of $\mathbb{F}_q$, but the complexity per iterations increases with q. This results from the fact that in each iteration all three algorithms have to solve a linear system over $\mathbb{F}_q$, and Lee and Brickell's and Stern's algorithms additionally iterate vectors over $\mathbb{F}_q$.

4.1   Prange's ISD Algorithm with Known Block Weights

For didactic reasons, we start with Prange's information-set decoder. Based on the knowledge of the Hamming weight of the error blocks, we adapt the strategy of choosing an information set, thereby (for known weight decomposition t) increasing the probability that the information set is error-free. For a given t, we fix a vector $x \in \{[x_1, \ldots, x_\ell] \in \mathbb{Z}^\ell : 0 \le x_i \le \eta_i - t_i, \sum_i x_i = k\}$. As described in the following, we have to choose x carefully since the work factor of our decoder will depend on it. Then, in each iteration of the algorithm, we choose independently and uniformly at random a subset $\mathcal{X}_i \subseteq \mathcal{W}_i$ of cardinality x_i for each i, i.e., x_i positions from each block. Since $\sum_i x_i = k$, the union $\mathcal{X} = \cup_i \mathcal{X}_i$ has cardinality k. Then, we proceed as in the original Prange algorithm: We check if $\mathcal{X}$ is an error-free information set.

Algorithm 1: Prange with Known Block Weights

Input: SDP (in form r and G, cf. discussion below Definition 1) and weight decomposition t of the error, vector x
Output: Error e
do

　　$\mathcal{X} \leftarrow \cup_{i=1}^{\ell} \mathcal{X}_i$, where the $\mathcal{X}_i$ are chosen independently uniformly from the subsets of $\mathcal{W}_i$ of cardinality x_i

　　$\boxed{e \leftarrow r - r_\mathcal{X}(G_\mathcal{X})^{-1} G \qquad \text{// Same as in original Prange alg.}}$

while $\mathrm{wt_H}(e) \neq \sum_i t_i$
return e

Theorem 7. *For a given $\boldsymbol{x}$, Algorithm 1 has expected work factor $W_{\mathsf{Prange}} = \frac{W_{\mathsf{Prange,Iter}}}{P_{\mathsf{Prange}}}$, where $W_{\mathsf{Prange,Iter}} = (n-k)^2(n+1)$ over $\mathbb{F}_2$ (cost of one iteration, see [28] for the cost over $\mathbb{F}_q$) and*

$$P_{\mathsf{Prange}} = \prod_{i=1}^{\ell} \binom{\eta_i - x_i}{t_i} \binom{\eta_i}{t_i}^{-1}$$

(success probability over any $\mathbb{F}_q$).

Proof. The cost of one iteration stays the same as in the original Prange decoder and is dominated by matrix inversion. An iteration succeeds if and only if the chosen positions in $\mathcal{X}$ are error-free[2]. This is again true iff every $\mathcal{X}_i$ is error-free. The fraction $\binom{\eta_i - x_i}{t_i}/\binom{\eta_i}{t_i}$ equals the probability that in block $\mathcal{W}_i$ of length η_i, the randomly chosen x_i positions are error-free, given that exactly t_i errors are contained in $\mathcal{W}_i$, i.e., exactly Prange's success probability restricted to one block. Since the positions $\mathcal{X}_i$ are chosen independently, the claim follows. □

Since $\boldsymbol{\eta}$ and $\boldsymbol{t}$ are given, we can influence the work factor of Algorithm 1 by choosing $\boldsymbol{x}$ in a suitable way. The following greedy algorithm maximizes the success probability for given $\boldsymbol{\eta}$ and $\boldsymbol{t}$:

– Initially, choose $x_i = \eta_i - t_i$ for all $i = 1, \dots, \ell$.
– While $\sum_i x_i > k$, decrease the x_j (by one) that increases P_{Prange} the most.

The greedy choice leads to a global maximum of P_{Prange} since the x_i only influence distinct factors of the product, so the increase of P_{Prange} resulting from decreasing one x_j does not influence the relative increase of P_{Prange} of other $x_i \neq x_j$ in the next steps. Note that the algorithm ensures that $x_i \leq \eta_i - t_i$ for all i, i.e., that the success probability P_{Prange} is non-zero. We will see in Sect. 4.4 that this choice of $\boldsymbol{x}$ reduces the work factor significantly compared to the original Prange algorithm for a growing number of blocks.

4.2 Lee–Brickell Algorithm with Known Block Weights

Lee and Brickell's algorithm works similarly to Prange's ISD algorithm. The difference is that the algorithm succeeds even if a few errors are contained in the randomly chosen information set. This means that the success probability is significantly increased compared to Prange's algorithm, but a bit more work is needed per iteration. We adapt the choice of the information sets in the Lee–Brickell algorithm: As in our adaptation of Prange's ISD, we choose from each block a given number of positions for the information set, and hope that overall at most a few errors are contained in the information set. The number of positions that we choose from each block depends on the distribution of errors. Algorithm 2 summarizes the decoder for blocks.

---

[2] As in most works on information-set decoding, we neglect the probability that a randomly chosen set is not an information set, since it is for most codes a constant in the same order of magnitude as 1.

Algorithm 2: Lee–Brickell with Known Block Weights

Input: SDP and weight decomposition t of the error, vector x, parameter p
Output: Error e
do

  $\mathcal{X} \leftarrow \cup_{i=1}^{\ell} \mathcal{X}_i$, where the $\mathcal{X}_i$ are chosen independently uniformly from the
  subsets of $\mathcal{W}_i$ of cardinality x_i

  $e \leftarrow$ Iteration of original Lee–Brickell alg. w.r.t. inf. set $\mathcal{X}$ and parameter p

while *Lee–Brickell stopping condition not satisfied for* e
return e

Theorem 8 states the expected work factor of the adapted Lee–Brickell algorithm. Note that the formula for the success probability in (1) consists of a sum whose number of summands may grow exponentially in the parameters p and ℓ. However, we can compute it in polynomial time using dynamic programming in a similar approach as in [30].

Theorem 8. *For a given x and parameter p, Algorithm 2 has an expected work factor of $W_{\text{LB}} = \frac{W_{\text{LB,Iter}}}{P_{\text{LB}}}$, where the cost per iteration is given by $W_{\text{LB,Iter}} := (n-k)^2(n+1) + (n-k)\sum_{i=1}^{p}\binom{k}{i} + (n-k)\binom{k}{p}$ over $\mathbb{F}_2$ (see [28] for the cost of one iteration over $\mathbb{F}_q$), and the success probability of each iteration is*

$$P_{\text{LB}} = \sum_{\substack{a \in \mathbb{Z}^{\ell} \\ 0 \leq a_i \leq t_i \\ \sum_i a_i = p}} \prod_{i=1}^{\ell} \frac{\binom{x_i}{a_i}\binom{\eta_i - x_i}{t_i - a_i}}{\binom{\eta_i}{t_i}}. \tag{1}$$

The success probability P_{LB} can be computed in polynomial bit complexity (in the parameters n, k, ℓ, p).

Proof. We adapt the original Lee–Brickell algorithm only by changing the selection of the information set. Hence, the cost per iteration $W_{\text{LB,Iter}}$ is the same as in the original algorithm. The cost is as stated if the concept of intermediate sums is used (cf. [8]).

The Lee–Brickell algorithm succeeds if the information set $\mathcal{X}$ is chosen such that it contains exactly p errors, and the remaining positions contain exactly $t - p$ errors. Hence, for P_{LB}, we sum over all possibilities that the p errors are distributed over the information set partitions $\mathcal{X}_i$ in the ℓ blocks. Obviously, there can only be at most t_i errors in the i-th information set. The products of fractions of binomial coefficients count the number of possibilities to choose information sets with $|\mathcal{X}_i| = x_i$ that contain exactly a_i errors, divided by the number of possibilities to distribute the t_i error positions on the η_i positions of a block.

We can compute P_{LB} in polynomial time using dynamic programming, see [30] for an efficient algorithm to compute a similar formula. □

Again, the success probability depends significantly on the choice of x for given η and t. A possible heuristic is to do the same as for Prange's decoder: Choose $x_i = \eta_i - t_i$ and decrease that x_i by one, for which (1) is increased the most. Since the best choice for p is often a small integer, it appears to be a good enough choice to use exactly the same x as computed for Prange's ISD ($p = 0$), even though p is chosen to be greater than 0.

4.3 Stern's ISD Algorithm with Known Block Weights

Stern's algorithm uses two parameters p and ν to choose an information set in each round. It allows a fixed number of errors in the information set and additionally restricts the number of errors outside the information set. Stern's algorithm divides the information set into two equal-size subsets $\mathcal{X}$ and $\mathcal{Y}$ and looks for words of weight p at the indices of $\mathcal{X}$ and weight p at the indices $\mathcal{Y}$, and weight 0 on a fixed uniform random set $\mathcal{Z}$ of ν positions outside the information set. Hence, we need to choose three sets, $\mathcal{X}, \mathcal{Y}, \mathcal{Z}$, at random. Again, we adapt the choice of these sets to the known weight distribution by designing three vectors x, y, z that indicate how many positions we choose for the three sets, respectively, from each block. A heuristic choice of the vectors is discussed below.

Algorithm 3: Stern with Known Block Weights

Input: SDP and weight decomposition t of the error, vectors x, y, z,
        parameters p, ν
Output: Error e
do

    $\mathcal{X} \leftarrow \cup_{i=1}^{\ell} \mathcal{X}_i$, where the $\mathcal{X}_i$ are chosen independently uniformly from the subsets of $\mathcal{W}_i$ of cardinality x_i
    $\mathcal{Y} \leftarrow \cup_{i=1}^{\ell} \mathcal{Y}_i$, where the $\mathcal{Y}_i$ are chosen independently uniformly from the subsets of $\mathcal{W}_i \setminus \mathcal{X}_i$ of cardinality y_i
    $\mathcal{Z} \leftarrow \cup_{i=1}^{\ell} \mathcal{Z}_i$, where the $\mathcal{Z}_i$ are chosen independently uniformly from the subsets of $\mathcal{W}_i \setminus (\mathcal{X}_i \cup \mathcal{Y}_i)$ of cardinality z_i

      | $e \leftarrow$ Iteration of original Stern alg. w.r.t. sets $\mathcal{X}, \mathcal{Y}, \mathcal{Z}$ and parameters p, ν |

while *Stern stopping condition not satisfied for e*
return e

Theorem 9 states the expected work factor of the adapted Stern algorithm. The formula for the success probability in (2) consists of sums whose number of summands may grow exponentially in the parameters p and ℓ. However, we can compute it in polynomial time using dynamic programming similar to [30].

Theorem 9. *For given x, y, z and parameters p, ν, Algorithm 3 has an expected work factor of $W_{\text{Stern}} := \frac{W_{\text{Stern,Iter}}}{P_{\text{Stern}}}$, where the cost per iteration over $\mathbb{F}_2$ (see [28] for the cost of one iteration over $\mathbb{F}_q$) is given by*

$$W_{\text{Stern,Iter}} = (n-k)^2(n+1) + \nu\left(L(m_x, p) + L(m_y, p) - m_x - m_y + \binom{m_y}{p}\right)$$

$$+ \frac{\binom{m_x}{p}\binom{m_y}{p}}{2^{\nu-1}}(t - 2p + 1)(2p + 1),$$

where $L(x, y) := \sum_{i=1}^{y} \binom{x}{i}$, $m_x := \sum_i x_i$, and $m_y := \sum_i y_i$.
The success probability of each iteration is given by

$$P_{\text{Stern}} = \sum_{\substack{a \in \mathbb{Z}^\ell \\ 0 \le a_i \le t_i \\ \sum_i a_i = p}} \sum_{\substack{b \in \mathbb{Z}^\ell \\ 0 \le b_i \le t_i - a_i \\ \sum_i b_i = p}} \prod_{i=1}^{\ell} \frac{\binom{x_i}{a_i}\binom{y_i}{b_i}\binom{\eta_i - x_i - y_i - z_i}{t_i - a_i - b_i}}{\binom{\eta_i}{t_i}}, \qquad (2)$$

and can be computed in polynomial bit complexity (in the parameters n, k, ℓ, p).

Proof. The proof works similar to the adapted Lee–Brickell algorithm in Subsect. 4.2. One iteration costs as much as in the original Stern algorithm, see [17] for a detailed analysis. Stern's stopping condition is fulfilled if and only if there are exactly p errors in $\mathcal{X}$, p errors in $\mathcal{Y}$, and 0 errors in $\mathcal{Z}$. The success probability formula then follows by summing over all cases to distribute p errors over the partition $\mathcal{X}_i$ of $\mathcal{X}$ (vector a) and to distribute p errors over the partition $\mathcal{Y}_i$ of $\mathcal{Y}$ (vector b). Again, the number of summands is exponential in n, k, ℓ, p, but we can compute P_{Stern} in polynomial time using dynamic programming. □

Again, the question is how to choose the vectors x, y, z for given η and t. We propose the following heuristic:

- Choose a vector $\tilde{x}$ with $\sum_i \tilde{x}_i = k$ as in the heuristic for Prange's decoder. Our goal is to choose an information set (union of $\mathcal{X}$ and $\mathcal{Y}$) with exactly $\tilde{x}_i$ positions from the i-th block.
- Choose $x_i, y_i \approx \frac{\tilde{x}_i}{2}$, e.g., alternatingly rounded down/up for odd $\tilde{x}_i$ such that we have $\sum_i x_i \approx \sum_i y_i \approx k/2$. The fact that we take roughly the same number of entries from the information subset in the i-th block means that the probability that $\mathcal{X}$ and $\mathcal{Y}$ contain exactly p errors is roughly the same.
- Choose $z_i = 0$ for all $i = 1, \ldots, \ell$. While $\sum_i z_i < \nu$, increase (by one) the z_j that maximizes P_{Stern}.

4.4 Numerical Results and Comparison

In Figs. 2, 3, 4 we present numerical results for the work factors of the modified Prange, Lee–Brickell, and Stern algorithms. All plots show logarithmic work factors as a function of the number of blocks ℓ. Figures 2, 3, and 4 contain the curves for the parameter sets of Classic McEliece, BIKE, and HQC, respectively, which we list in the preliminaries.

We use the presented heuristics to choose x, y, z, and optimize over the parameters p and ν. If $\ell \nmid n$, we choose $\eta_i \approx n/\ell$ rounded up or down in a suitable ratio. Since the work factors depend heavily on the weight distribution to the blocks, we randomly sample for each parameter set several errors (uniformly at random from the set of errors of weight t) and present realizations as points, plus curves for the log of the mean work factor, taken over the realizations. The number of realizations for each ℓ and algorithm is roughly 50.

It can be seen that for all systems, parameter sets, and algorithms, only a few blocks are needed to push the work factor below the claimed security level. For instance, in `mceliece348864`, Stern's algorithm only needs to know the weight

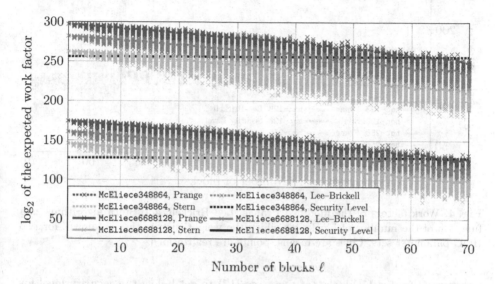

Fig. 2. Work factors of the modified Prange, Lee–Brickell, and Stern algorithms as a function of the number of blocks for which the weight distribution is known, for two Classic McEliece parameter sets. Lines are means, points are realizations.

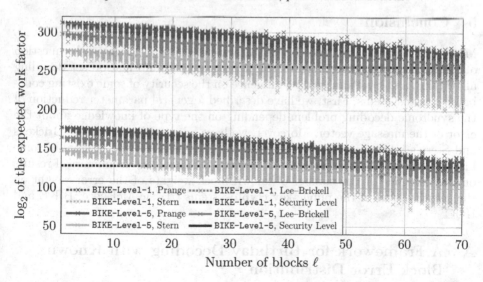

Fig. 3. Work factors of the modified Prange, Lee–Brickell, and Stern algorithms as a function of the number of blocks for which the weight distribution is known, for two BIKE parameter sets. Lines are means, points are realizations.

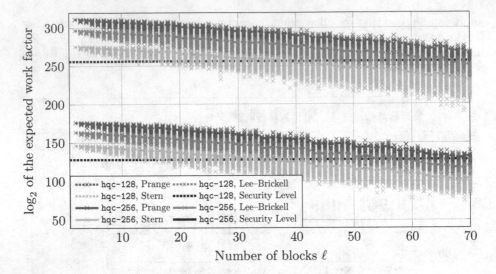

Fig. 4. Work factors of the modified Prange, Lee–Brickell, and Stern algorithms as a function of the number of blocks for which the weight distribution is known, for two HQC parameter sets. Lines are means, points are realizations.

decomposition for 11 blocks (of size $\eta_i \approx 317$) to get below the security level for *some* weight decompositions (i.e., realizations of the error), and 29 blocks (of size $\eta_i \approx 120$) to push the *mean* work factor below the claimed security level.

5   Conclusion

Motivated by (template) side-channel attacks, we have investigated the question of how side information improves the performance of information set decoding algorithms, and have analyzed its influence on the security of some existing code-based cryptosystems. First, we have described a general parameter reduction of the syndrome decoding problem depending on the type of knowledge about the error or the message vector. Moreover, we have adapted Prange's, Lee-Brickel's and Stern's information set decoders to the setting where the weights of some blocks of the error vector are known. Finally, we have derived numerical results on the influence of the number of known block weights of the error weight on the security level of the public-key cryptosystems Classic McEliece, BIKE and HQC.

A   A Framework for Birthday Decoding with Known Block Error Distribution

As an outlook for future work, we describe a general framework for generalized birthday decoders, for the case that we know the error distribution over prescribed blocks of the error vector. For this we will use the "Partial Gaussian

Elimination (PGE) + Small SDP" setup. In the "Small SDP" step several versions of birthday decoding can be used, see e.g. [6,21], however we will restrict ourselves to Wagner's original idea from [41].

As in Sect. 4 consider an SDP instance with parity check matrix $H \in \mathbb{F}_2^{(n-k)\times n}$ and syndrome $s \in \mathbb{F}_2^{n-k}$. Assume we have a partition of $\{1,\ldots,n\}$ into $\mathcal{W}_1,\ldots,\mathcal{W}_\ell$ and $[t_1,\ldots,t_\ell]$ such that $t_i = \mathrm{wt_H}(e_{\mathcal{W}_i})$ and $\sum_{i=1}^\ell t_i = t$.

We proceed as follows:

1. <u>Partial Gaussian Elimination:</u>
   Choose some $v \leq n-k$ and subsets $\mathcal{X}_i \subseteq \mathcal{W}_i$ such that for $x_i := |\mathcal{X}_i|$ we have $\sum_i x_i = k+v$. Find an $n \times n$ permutation matrix P that moves the columns of H indexed by the $\mathcal{X}_i$ to the left[3], and find and invertible $U \in \mathbb{F}_2^{(n-k)\times(n-k)}$ such that
   $$UHP = \begin{pmatrix} A & I_{n-k-v} \\ B & 0 \end{pmatrix}.$$
   Denote $[s_1\ s_2] := sU^\top$ and $eP := [e_1\ e_2]$, then the original syndrome equation $eH^\top = s$ is equivalent to $eP(P^\top H^\top U^\top) = sU^\top$ and splits into
   $$e_1 A^\top + e_2 = s_1$$
   $$e_1 B^\top = s_2$$

   Prescribe an error weight $p \leq t$ for e_1, with which the second equation is now a smaller SDP instance, with possibly more than one solution. Find these solutions as explained in the step below, and for each of those check if the first equation above gives a valid solution, i.e., if $\mathrm{wt_H}(s_1 - e_1 A^\top) = t-p$. If so, then this is a solution to the original SDP.
2. <u>Small SDP with Wagner:</u>
   We choose a number b of levels for this step, and partition $\mathcal{X} = \bigcup_{i=1}^\ell \mathcal{X}_i$ into 2^b subsets of size v_i, respectively, such that $\sum_{i=1}^{2^b} v_i = \sum_{i=1}^\ell x_i = k+v$. Then we distribute the p errors over the corresponding entries of e_1, assuming that p_i errors happened in the i-th subset. Here the first open question arises, namely how to choose these sets and the error distribution. In the classical setup you can simply choose (random) subsets of the same size, and assume that each of the corresponding sets of coordinates in e_1 has weight $\frac{p}{2^b}$. However, in our setting this is most likely a suboptimal choice, and the subsets should be chosen with respect to the block partition and the error distribution.
   In particular, if $\ell = 2^b$, then we can choose the 2^b subsets equal to $\mathcal{X}_1,\ldots,\mathcal{X}_\ell$, and assume that $p_i \leq t_i$ errors are spread over the coordinates of e indexed by $\mathcal{X}_i$, such that $\sum_{i=1}^\ell p_i = p$.
   Next we build the initial lists of vectors in $\mathbb{F}_2^{v_i}$ of weight p_i, for $i = 1,\ldots,2^b$, respectively. Then we proceed with the merge-concatenate operation on the b levels as in the classical setting (where for each level, we can choose $u_1 < u_2 < \cdots < u_b = k+v$, which indicates how many coordinates we want to

---
[3] This step is technically not necessary, but it simplifies the notation below.

merge on), until we have a final list of candidates for e_1, solving the small SDP instance.

As an example, if we consider the simplest case with $b = 1$ and $\ell = 2^b = 2$, we construct two lists

$$\mathcal{L}_1^{(0)} := \{v \in \mathbb{F}_2^{x_1} \mid \mathrm{wt}_H(v) = p_1\}$$

$$\mathcal{L}_2^{(0)} := \{v \in \mathbb{F}_2^{x_2} \mid \mathrm{wt}_H(v) = p_2\}$$

such that $x_1 + x_2 = k + v$ and $p_1 + p_2 = p$.[4] Let us write $B = [B_1 \ B_2]$, such that $B_i \in \mathbb{F}_2^{v \times x_i}$. Then we merge-concatenate to produce the final list

$$\mathcal{L}^{(1)} := \{(v_1, v_2) \in \mathcal{L}_1^{(0)} \times \mathcal{L}_2^{(0)} \mid v_1 B_1^\top = s_2 - v_2 B_2^\top\},$$

which consists of solutions to the small SDP, i.e., candidates for e_1.

Finally, we will give some idea about the computational complexity of this algorithm:

1. The success probability depends on the assumed error distribution and the choices of the free parameters. Similarly to the Lee-Brickell adaptation, it is

$$\sum_{\substack{a \in \mathbb{Z}^\ell \\ 0 \le a_i \le t_i \\ \sum_i a_i = p}} \prod_{i=1}^{\ell} \frac{\binom{x_i}{a_i}\binom{\eta_i - x_i}{t_i - a_i}}{\binom{\eta_i}{t_i}} \cdot P_{\mathrm{merge}}(u_1, \dots, u_{b-1}),$$

where $P_{\mathrm{merge}}(u_1, \dots, u_{b-1})$ is the probability that the sought-after solution has zeros in the coordinates prescribed by $u_1, \dots, u_{b-1}$.[5]

2. Each iteration depends on the number of levels, the u_i and the size of the original lists, and is comparable to the cost of one iteration of the algorithm in the classical setting (assuming that the initial lists are of the same size).

The speed-up compared to not using the knowledge about the error distribution in the blocks of the error vector is hence mainly in the success probability. This is analogous to the algorithms studied in Sect. 4.

References

1. Aguilar-Melchor, C., et al.: Hamming Quasi-Cyclic (HQC), Third round submission to the NIST post-quantum cryptography call (2019). https://pqc-hqc.org
2. Albrecht, M.R., et al.: Classic McEliece, Third round submission to the NIST post-quantum cryptography call (2019). https://classic.mceliece.org

---

[4] To make proper use of the birthday paradox we could also take random subsets of the above lists, instead of all vectors of prescribed length and weight.

[5] If we take only subsets of the initial lists, then this will also be considered in this success probability.

3. Aragon, N., et al.: BIKE: Bit Flipping Key Encapsulation, Third round submission to the NIST post-quantum cryptography call (2019). https://bikesuite.org/

4. Augot, D., Finiasz, M., Sendrier, N.: A fast provably secure cryptographic hash function, Cryptology ePrint Archive, Report 2003/230 (2003). https://eprint.iacr.org/2003/230

5. Baldi, M., et al.: A new path to code-based signatures via identification schemes with restricted errors (2020)

6. Becker, A., Joux, A., May, A., Meurer, A.: Decoding random binary linear codes in $2^{n/20}$: How $1 + 1 = 0$ improves information set decoding. In: Pointcheval, D., Johansson, T. (eds.) EUROCRYPT 2012. LNCS, vol. 7237, pp. 520–536. Springer, Heidelberg (2012). https://doi.org/10.1007/978-3-642-29011-4_31

7. Berlekamp, E., McEliece, R., van Tilborg, H.: On the inherent intractability of certain coding problems (corresp.). IEEE Trans. Inf. Theor. **24**(3), 384–386 (1978)

8. Bernstein, D.J., Lange, T., Peters, C.: Smaller decoding exponents: ball-collision decoding. In: Rogaway, P. (ed.) CRYPTO 2011. LNCS, vol. 6841, pp. 743–760. Springer, Heidelberg (2011). https://doi.org/10.1007/978-3-642-22792-9_42

9. Bernstein, D.J., Lange, T., Peters, C., Schwabe, P.: Faster 2-regular information-set decoding. In: Chee, Y.M., Guo, Z., Ling, S., Shao, F., Tang, Y., Wang, H., Xing, C. (eds.) IWCC 2011. LNCS, vol. 6639, pp. 81–98. Springer, Heidelberg (2011). https://doi.org/10.1007/978-3-642-20901-7_5

10. Canteaut, A.: A new algorithm for finding minimum-weight words in a linear code: Application to McEliece's cryptosystem and to narrow-sense bch codes of length 511. IEEE Trans. Inf. Theor. **44**, 367–378 (1998)

11. Chou, T.: QcBits: constant time small-key code-based cryptography. In: Gierlichs, B., Poschmann, A.Y. (eds.) CHES 2016. LNCS, vol. 9813, pp. 280–300. Springer, Heidelberg (2016). https://doi.org/10.1007/978-3-662-53140-2_14

12. Choudary, M.O., Kuhn, M.G.: Efficient, portable template attacks. IEEE Trans. Inf. Forensics Secur. **13**(2), 490–501 (2018)

13. Dachman-Soled, D., Ducas, L., Gong, H., Rossi, M.: LWE with side information: attacks and concrete security estimation. In: Micciancio, D., Ristenpart, T. (eds.) CRYPTO 2020. LNCS, vol. 12171, pp. 329–358. Springer, Cham (2020). https://doi.org/10.1007/978-3-030-56880-1_12

14. Diffie, W., Hellman, M.: New directions in cryptography. IEEE Trans. Inf. Theor. **22**(6), 644–654 (1976)

15. Gligoroski, D., Samardjiska, S., Jacobsen, H., Bezzateev, S.: McEliece in the world of Escher, Cryptology ePrint Archive, Report 2014/360 (2014)

16. Heyse, S., Moradi, A., Paar, C.: Practical power analysis attacks on software implementations of McEliece. In: Sendrier, N. (ed.) PQCrypto 2010. LNCS, vol. 6061, pp. 108–125. Springer, Heidelberg (2010). https://doi.org/10.1007/978-3-642-12929-2_9

17. Horlemann-Trautmann, A.-L., Weger, V.: Information set decoding in the Lee metric with applications to cryptography. Adv. Math. Commun. (2020)

18. Interlando, C., Khathuria, K., Rohrer, N., Rosenthal, J., Weger, V.: Generalization of the ball-collision algorithm. J. Algebra Comb. Discrete Struct. Appl. **7**, 195–207 (2020)

19. Lahr, N., Niederhagen, R., Petri, R., Samardjiska, S.: Side channel information set decoding using iterative chunking. In: Moriai, S., Wang, H. (eds.) ASIACRYPT 2020. LNCS, vol. 12491, pp. 881–910. Springer, Cham (2020). https://doi.org/10.1007/978-3-030-64837-4_29

20. Lee, P.J., Brickell, E.F.: An observation on the security of McEliece's public-key cryptosystem. In: Barstow, D., et al. (eds.) EUROCRYPT 1988. LNCS, vol. 330, pp. 275–280. Springer, Heidelberg (1988). https://doi.org/10.1007/3-540-45961-8_25

21. May, A., Meurer, A., Thomae, E.: Decoding random linear codes in $\tilde{\mathcal{O}}(2^{0.054n})$. In: Lee, D.H., Wang, X. (eds.) ASIACRYPT 2011. LNCS, vol. 7073, pp. 107–124. Springer, Heidelberg (2011). https://doi.org/10.1007/978-3-642-25385-0_6

22. McEliece, R.J.: A public-key cryptosystem based on algebraic coding theory. DSN Prog. Rep. **44**, 114–116 (1978)

23. Molter, H.G., Stöttinger, M., Shoufan, A., Strenzke, F.: A simple power analysis attack on a McEliece cryptoprocessor. J. Cryptogr. Eng. **1**(1), 29–36 (2011). https://doi.org/10.1007/s13389-011-0001-3

24. Moody, D., Perlner, R.: Vulnerabilities of "McEliece in the world of Escher". In: Takagi, T. (ed.) PQCrypto 2016. LNCS, vol. 9606, pp. 104–117. Springer, Cham (2016). https://doi.org/10.1007/978-3-319-29360-8_8

25. National Institute of Standards and Technology (NIST), U.S. Department of Commerce, Post-quantum cryptography standardization (2017). https://csrc.nist.gov/Projects/post-quantum-cryptography/Post-Quantum-Cryptography-Standardization

26. Niebuhr, R., Persichetti, E., Cayrel, P.-L., Bulygin, S., Buchmann, J.: On lower bounds for information set decoding over F_q and on the effect of partial knowledge. Int. J. Inf. Coding Theor. **4**(1), 47–78 (2017)

27. Paiva, T.B., Terada, R.: A timing attack on the HQC encryption scheme. In: Paterson, K.G., Stebila, D. (eds.) SAC 2019. LNCS, vol. 11959, pp. 551–573. Springer, Cham (2020). https://doi.org/10.1007/978-3-030-38471-5_22

28. Peters, C.: Information-set decoding for linear codes over F_q. In: Sendrier, N. (ed.) PQCrypto 2010. LNCS, vol. 6061, pp. 81–94. Springer, Heidelberg (2010). https://doi.org/10.1007/978-3-642-12929-2_7

29. Prange, E.: The use of information sets in decoding cyclic codes. IRE Trans. Inf. Theor. **8**(5), 5–9 (1962)

30. Puchinger, S., Renner, J., Rosenkilde, J.: Generic decoding in the sum-rank metric. In: IEEE International Symposium on Information Theory (ISIT), Extended Version, pp. 54–59 (2020) arxiv: https://arxiv.org/abs/2001.04812

31. Rossi, M., Hamburg, M., Hutter, M., Marson, M.E.: A side-channel assisted cryptanalytic attack against QcBits. In: Fischer, W., Homma, N. (eds.) CHES 2017. LNCS, vol. 10529, pp. 3–23. Springer, Cham (2017). https://doi.org/10.1007/978-3-319-66787-4_1

32. Schamberger, T., Renner, J., Sigl, G., Wachter-Zeh, A.: A power side-channel attack on the CCA2-secure HQC KEM, Cryptology ePrint Archive, Report 2020/910 (2020). https://eprint.iacr.org/2020/910

33. Shor, P.W.: Algorithms for quantum computation: discrete logarithms and factoring. In: Proceedings 35th Annual Symposium on Foundations of Computer Science, pp. 124–134 (1994)

34. Shoufan, A., Strenzke, F., Molter, H.G., Stöttinger, M.: A timing attack against Patterson algorithm in the McEliece PKC. In: Lee, D., Hong, S. (eds.) ICISC 2009. LNCS, vol. 5984, pp. 161–175. Springer, Heidelberg (2010). https://doi.org/10.1007/978-3-642-14423-3_12

35. Sim, B.-Y., Kwon, J., Choi, K.Y., Cho, J., Park, A., Han, D.-G.: Novel side-channel attacks on quasi-cyclic code-based cryptography. IACR Trans. Cryptographic Hardware Embed. Syst. 2019(4) (2019)

36. Stern, J.: A method for finding codewords of small weight. In: Cohen, G., Wolf-mann, J. (eds.) Coding Theory 1988. LNCS, vol. 388, pp. 106–113. Springer, Heidelberg (1989). https://doi.org/10.1007/BFb0019850
37. Strenzke, F.: Timing attacks against the syndrome inversion in code-based cryptosystems. In: Gaborit, P. (ed.) PQCrypto 2013. LNCS, vol. 7932, pp. 217–230. Springer, Heidelberg (2013). https://doi.org/10.1007/978-3-642-38616-9_15
38. Strenzke, F., Tews, E., Molter, H.G., Overbeck, R., Shoufan, A.: Side channels in the McEliece PKC. In: Buchmann, J., Ding, J. (eds.) PQCrypto 2008. LNCS, vol. 5299, pp. 216–229. Springer, Heidelberg (2008). https://doi.org/10.1007/978-3-540-88403-3_15
39. von Maurich, I., Güneysu, T.: Towards side-channel resistant implementations of QC-MDPC McEliece encryption on constrained devices. In: Mosca, M. (ed.) PQCrypto 2014. LNCS, vol. 8772, pp. 266–282. Springer, Cham (2014). https://doi.org/10.1007/978-3-319-11659-4_16
40. Wafo-Tapa, G., Bettaieb, S., Bidoux, L., Gaborit, P., Marcatel, E.: A practicable timing attack against HQC and its countermeasure, Cryptology ePrint Archive, Report 2019/909 (2019). https://eprint.iacr.org/2019/909
41. Wagner, D.: A generalized birthday problem. In: Yung, M. (ed.) CRYPTO 2002. LNCS, vol. 2442, pp. 288–304. Springer, Heidelberg (2002). https://doi.org/10.1007/3-540-45708-9_19

A Correction to a Code-Based Blind Signature Scheme

Olivier Blazy[(✉)], Philippe Gaborit[(✉)], and Dang Truong Mac[(✉)]

University of Limoges, Limoges, France
{olivier.blazy,gaborit}@unilim.fr, dang.mac@etu.unilim.fr

Abstract. This work proposes a reparation to the flaw in the paper of Blazy *et al.* (IEEE 2017). The flaw lies in the proof of the unforgeability property. More precisely, the way of handling collisions and of using the adversary to solve the challenge problem are incorrect. This problem is circumvented by adding a proof of knowledge of the randomness. It results in a scheme with the same public key size as that of the previous one, the size of the signature is a little bit larger.

Keywords: Zero-knowledge protocols · Stern's SD scheme · CFS signature · Code-based cryptography

1 Introduction

Blind signatures were first introduced by D. Chaum in 1982 [5]. Unlike usual signatures, the signed message is hidden from the signer (blindness). With this property, blind signatures have found many applications such as electronic voting, electronic cash [5]. There has been many blind signature protocols most of which use the RSA approach [15,17] and thus are not considered to be post-quantum secure. The task of constructing post-quantum secure blind signature schemes was first successfully handled by Hauck *et al.* [13]. Their construction is based on lattice assumptions aided by linear hash functions.

In the code-based field, blind signatures were first considered by Overbeck [16]. Still, the construction has many issues to be reflected on. The second notable attempt was made in 2017, a code-based blind signature scheme was proposed by Blazy *et al.* [4]. However, there is a flaw in the proof of the unforgeability property due to a shortcoming in the construction. Thus, the proof therein is invalid. The goal of this short paper is to give a new blind signature scheme based on the one in [4], which is supported by correct proofs of security.

Organization. The rest of the paper is organized as follows. In Sect. 2, we briefly recall some basic notions in code-based cryptography, which are required for our construction. Section 3 describes the security model for our scheme. Section 4 recalls the previous scheme and provides the explanation of the flaw in its proof. A new scheme having correct proofs of security and a set of parameters are presented in Sect. 5. Finally, we draw some remarks in Sect. 6.

ⓒ Springer Nature Switzerland AG 2022
A. Wachter-Zeh et al. (Eds.): CBCrypto 2021, LNCS 13150, pp. 84–94, 2022.
https://doi.org/10.1007/978-3-030-98365-9_5

2   Background on Code-Based Cryptography

Let $\mathbb{F}_q$ denote the finite field with q elements, $\mathbf{H} \in \mathbb{F}_q^{(n-k)\times n}$ a parity-check matrix of some linear code of length n and dimension k, and $\mathcal{S}_w^n$ the sphere of $\mathbb{F}_q^n$ of radius w. Throughout the paper, $h(\cdot)$ will stand for a cryptographic hash function.

2.1   Syndrome Decoding

We recall some well-known notions as well as hardness assumptions related to syndrome decoding.

Definition 1 (Syndrome Decoding Distribution). *Let n, k, and w be positive integers. The syndrome decoding distribution, denoted by $\mathsf{SD}(n, k, w)$, chooses $\mathbf{H} \leftarrow \mathbb{F}_q^{(n-k)\times n}$ and $\mathbf{x} \leftarrow \mathcal{S}_w^n$, and outputs $\left(\mathbf{H}, \mathbf{H}\cdot\mathbf{x}^T\right)$.*

Definition 2 (Computational Syndrome Decoding Problem). *Given a random matrix $\mathbf{H} \in \mathbb{F}_q^{(n-k)\times n}$, a word $\mathbf{u} \in \mathbb{F}_q^{n-k}$, and an integer $w > 0$, the computational syndrome decoding $\mathsf{CSD}(\mathbf{H}, \mathbf{u}, w)$ asks to find a vector $\mathbf{x} \in \mathbb{F}_q^n$ of Hamming weight at most w such that $\mathbf{H}\cdot\mathbf{x}^T = \mathbf{u}^T$.*

The decisional version of the above problem is stated as follows

Definition 3 (Decisional Syndrome Decoding Problem). *The decision SD problem, denoted by $\mathsf{DSD}(n, k, w)$, asks to decide with non-negligible advantage whether an instance $(\mathbf{H}, \mathbf{y}^T)$ came from the $\mathsf{SD}(n, k, w)$ distribution or the uniform distribution over $\mathbb{F}_q^{(n-k)\times n} \times \mathbb{F}_q^{n-k}$.*

One of the most important parameters in code-based cryptography is the Gilbert-Varshamov Bound, it is defined as follows.

Definition 4. *The volume of a ball of radius w in the Hamming space $\mathbb{F}_q^n$ is*

$$V_n(w) = \sum_{i=0}^{w}(q-1)^i\binom{n}{i}.$$

For given n and k, the smallest integer b_{GV} such that $V_n(b_{GV}) \geq q^{n-k}$ is called the Gilbert-Varshamov (GV) bound.

Definition 5. *We call w-bounded decoder associated to $\mathbf{H}$ a procedure $\mathbb{F}_q^{n-k} \rightarrow \mathbb{F}_q^n$ which returns for all $\mathbf{u} \in \mathbb{F}_q^{n-k}$ an element of $\mathsf{CSD}(\mathbf{H}, \mathbf{u}, w)$ (or fails if this set is empty).*

For given n and k and for almost all codes, a w-bounded decoder fails for a proportion approximately $\exp(-V_n(w)/q^{n-k})$ of the instances. If we choose an integer $w > b_{GV}$, a w-bounded decoder almost never fails[1]. We will speak of a *complete*[2] decoder.

---

[1] Most of the time $w = b_{GV}$ is enough, exceptionally $w = b_{GV} + 1$.

[2] The word *complete* is used here for convenience, the decoder may fail but for a negligible proportion of the instances.

2.2 Trapdoor Digital Signature

Let w_0 be the smallest integer such that $\mathsf{CSD}(\mathbf{H}, \mathbf{u}, w_0) \neq \varnothing$ with high probability (*i.e.*, from the previous section $w_0 = \lceil b_{gv} \rceil$ or $\lceil b_{gv} + 1 \rceil$). We assume here that the linear code defined by the parity check matrix $\mathbf{H}$ has some hidden algebraic structure (for instance a binary Goppa code) which enables a trapdoor complete w_0-bounded decoder $D_{\mathbf{H}}(\cdot)$.

CFS Signatures. Obtaining a practical complete decoder is not an easy task because the desired decoding bound w_0 is above the algebraic error correcting capability. It is possible for binary Goppa codes of high rate (*i.e.*, the ratio k/n between dimension and length is close to 1) [6]: the resulting complete decoder is complex but still has an exponential advantage in complexity compared with the best generic algorithms for solving CSD.

Let $\mathbf{H}$ be the parity check matrix of a CFS code, let $D_{\mathbf{H}}(\cdot)$ be the trapdoor CFS decoding function. The CFS problem is defined as given q accesses to a CFS oracle (given $\mathbf{x}$, it returns $\mathbf{y} = D_{\mathbf{H}}(\mathbf{x})$), and $\mathbf{u}^*$ the adversary has to return $\mathbf{y}^*$ such that $\mathbf{H}\mathbf{y}^{*T} = \mathbf{u}^{*T}$ and $wt(\mathbf{y}^*) = w$ in polynomial time after at most q queries to the oracle, on words different from $\mathbf{u}^*$.

Fig. 1. The CFS problem.

Security of CFS Signatures. Parity-check matrices of high rate Goppa codes can be distinguished from random matrices [8]. Still, this distinguishing attack does not lead to an efficient key recovery attack (recovering $D_{\mathbf{H}}$ from $\mathbf{H}$), however it invalidates the security reduction given in [6]. We refer to [14] for more details on the security of CFS.

Parallel CFS. A modification of the CFS scheme was proposed by Finiasz [10]. It consists in producing λ signatures (3 or 4) of related digests. If done correctly, the cost for an existential forgery attack can be made arbitrarily close to the cost for a universal forgery attack.

2.3 Stern's Authentication Protocol

This section is dedicated to Stern's authentication protocols and, in particular, to a concatenated Stern authentication protocol. The latter is the randomized version of the one given in [1] and will serve as a building-block in our construction.

In 1993, J. Stern proposed an authentication algorithm in [21]. In the protocol, the prover $\mathcal{P}$ convinces the verifier $\mathcal{V}$ that he knows a secret word $\mathbf{s} \in \mathbb{F}_2^n$ of weight w such that $\mathbf{u}^T = \mathbf{H} \cdot \mathbf{s}^T$, where $\mathbf{u} \in \mathbb{F}_2^{n-k}$ and $\mathbf{H} \in \mathbb{F}_2^{(n-k) \times n}$ are public. A fake prover has a probability $2/3$ to cheat at each iteration and thus many iterations are needed (137 iterations for a cheating probability $< 2^{-80}$). Later on, this construction was usually combined with the Fiat-Shamir paradigm [9] to produce digital signatures, and hence, together, they become an essential part of many schemes. In the next paragraph, we discuss a variation of the original Stern's protocol, namely, the concatenated Stern authentication protocol.

For ease of notations, from now on, k (and k' in the next sections) will have the equal meaning of co-dimension. Let us consider $\mathbf{Q}$ a $k \times n_1$ binary matrix and $\mathbf{R}$ a $k \times n_2$ binary matrix. Suppose that there exist a vector $(\mathbf{x}, \mathbf{y})$ with $\mathbf{x}, \mathbf{y}$ of respective lengths n_1, n_2 and of weight w_1, w_2, and a syndrome $\mathbf{s}$ such that $[\mathbf{Q}|\mathbf{R}] \cdot (\mathbf{x}, \mathbf{y})^T = \mathbf{s}^T = \mathbf{Q} \cdot \mathbf{x}^T + \mathbf{R} \cdot \mathbf{y}^T$. The (randomized) concatenated Stern authentication protocol is a zero-knowledge (ZK) protocol which allows the prover $\mathcal{P}$ to prove that he knows a vector $(\mathbf{x}, \mathbf{y})$, for $\mathbf{x}$ and $\mathbf{y}$ of respective weight w_1 and w_2 such that $[\mathbf{Q}|\mathbf{R}] \cdot (\mathbf{x}, \mathbf{y})^T = \mathbf{s}^T = \mathbf{Q} \cdot \mathbf{x}^T + \mathbf{R} \cdot \mathbf{y}^T$. In the following, S_n denotes the permutation group of length n, and $|$ stands for concatenation. The protocol works as described in Fig. 2.

Concatenated Stern zero-knowledge authentification protocol
Public data: two matrices $\mathbf{Q}$ and $\mathbf{R}$ of respective size $k \times n_1$ and $k \times n_2$, a syndrome $\mathbf{s}$.
Prover $\mathcal{P}$: a vector $(\mathbf{x}, \mathbf{y})$ for $\mathbf{x}$ and $\mathbf{y}$ of respective weight w_1 and w_2 such that

$$[\mathbf{Q}|\mathbf{R}] \cdot (\mathbf{x}, \mathbf{y})^T - \mathbf{s}^T = \mathbf{Q} \cdot \mathbf{x}^T + \mathbf{R} \cdot \mathbf{y}^T.$$

The prover $\mathcal{P}$ interacts with a verifier $\mathcal{V}$ in 3 steps and a verification:

1. **Commitments.** $\mathcal{P}$ generates $\sigma_1 \leftarrow S_{n_1}$, $\sigma_2 \leftarrow S_{n_2}$, $\mathbf{u}_1 \leftarrow \mathbb{F}_2^{n_1}, \mathbf{u}_2 \leftarrow \mathbb{F}_2^{n_2}$ and $r_1, r_2, r_3 \leftarrow 1^\lambda$.
   $\mathcal{P}$ sends three commitments:
   $c_1 = h(\sigma_1(\mathbf{u}_1)|\sigma_2(\mathbf{u}_2)|r_1)$,
   $c_2 = h(\sigma_1|\mathbf{Q}\mathbf{u}_1^T + \mathbf{R}\mathbf{u}_2^T|\sigma_2|r_2)$,
   $c_3 = h(\sigma_1(\mathbf{x}+\mathbf{u}_1)|\sigma_2(\mathbf{y}+\mathbf{u}_2)|r_3)$.
2. **Challenge.** $\mathcal{V}$ responds with $b \in \{0, 1, 2\}$.
3. **Answer.** There are three possibilities:
   - If $b = 0$, P reveals $\sigma_1(\mathbf{u}_1)$, $\sigma_2(\mathbf{u}_2)$, $\sigma_1(\mathbf{x})$, $\sigma_2(\mathbf{y})$, r_1, and r_3.
   - If $b = 1$, he reveals σ_1, σ_2, $\mathbf{x}+\mathbf{u}_1$, $\mathbf{y}+\mathbf{u}_2$, r_2, and r_3.
   - If $b = 2$, he reveals σ_1, $\mathbf{u}_1$, σ_2, $\mathbf{u}_2$, r_1, and r_2.
4. **Verification.**
   - if $b = 0$ checks c_1 and c_3;
   - if $b = 1$ checks c_2 and c_3;
   - if $b = 2$ checks c_1 and c_2.

Fig. 2. Concatenated Stern zero-knowledge protocol.

Theorem 1. *The concatenated Stern zero-knowledge protocol is a ZK protocol with cheating probability $\frac{2}{3}$.*

Proof. The protocol we describe is an adaptation of the protocol described in [1] to which we added random values r_1, r_2 and r_3. By doing so, the protocol cannot be testable and leaks no information (see [1] for details on testable Stern protocol). For verification, the only non trivial check is (as for Stern's original protocol) for $b = 1$ and the value c_1, which is checked with the public syndrome $\mathbf{s}$, one recovers $\mathbf{Qu}_1^T + \mathbf{Ru}_2^T$ as $\mathbf{Qu}_1^T + \mathbf{Ru}_2^T = \mathbf{Q}(\mathbf{x} + \mathbf{u}_1)^T + \mathbf{R}(\mathbf{y} + \mathbf{u}_2)^T - \mathbf{s}^T$. The proof is straightforward from [1] with the ZK properties, argued by using the randomness of r_1, r_2, and r_3.   □

3   Blind Signature

As formalized by Pointcheval and Stern [18], a blind signature scheme involves two parties, a user $\mathcal{U}$ and a signer $\mathcal{S}$. The user submits a masked (or blinded) message that the signer will sign with a digital signature scheme whose public key is known. This part is named BSProtocol. The user unmasks this signature to build a signature of the unmasked message which is valid for the signer's public key. A verification can be made on the final signature with the signer's public key.

More precisely, we can derive the definition of blind signatures from that of digital signatures. Instead of having a signing phase $\mathsf{Sign}(\mathsf{sk}, M; \mu)$, we have an interactive phase $\mathsf{BSProtocol}\langle \mathcal{S}, \mathcal{U}\rangle$ between the user $\mathcal{U}(\mathsf{vk}, M; \rho)$, who will (probably) transmit a masked information on M under some randomness ρ in order to obtain a signature valid under the verification key vk, and the signer $\mathcal{S}(\mathsf{sk}; \mu)$, who will generate something based on this value, and his secret key which should lead the user to a valid signature. Such signatures are correct if when both the user and signer are honest then $\mathsf{BSProtocol}\langle \mathcal{S}, \mathcal{U}\rangle$ does indeed lead to valid signature on M under vk. There are two additional security properties, one protecting the signer, the other the user.

- On one hand, there is an *Unforgeability* property, which states that a malicious user should not be able to compute $n+1$ valid signatures on different messages after at most n interactions with the signer.
- On the other hand, the *Blindness* property says that a malicious signer who signed two messages M_0 and M_1 should not be able to decide which one was signed first.

These properties are described in Fig. 3.

$\mathsf{Exp}^{\mathsf{bl}-b}_{BS,\mathcal{S}^*}(\mathfrak{K})$
1. $(\mathsf{param}) \leftarrow \mathsf{BSSetup}(1^{\mathfrak{K}})$
2. $(\mathsf{vk}, M_0, M_1) \leftarrow \mathcal{A}(\mathtt{FIND} : \mathsf{param})$
3. $\sigma_b \leftarrow \mathsf{BSProtocol}\langle \mathcal{A}, \mathcal{U}(\mathsf{vk}, M_b)\rangle$
4. $\sigma_{1-b} \leftarrow \mathsf{BSProtocol}\langle \mathcal{A}, \mathcal{U}(\mathsf{vk}, M_{1-b})\rangle$
5. $b^* \leftarrow \mathcal{S}^*(\mathtt{GUESS} : M_0, M_1)$;
6. $\mathtt{RETURN}\ b^* = b$.

$\mathsf{Exp}^{\mathsf{uf}}_{BS,\mathcal{U}^*}(\mathfrak{K})$
1. $(\mathsf{param}) \leftarrow \mathsf{BSSetup}(1^{\mathfrak{K}})$
2. $(\mathsf{vk}, \mathsf{sk}) \leftarrow \mathsf{BSKeyGen}(\mathsf{param})$
3. For $i = 1, \ldots, q_s$, $\mathsf{BSProtocol}\langle \mathcal{S}(\mathsf{sk}), \mathcal{A}(\mathtt{INIT} : \mathsf{vk})\rangle$
4. $((m_1, \sigma_1), \ldots, (m_{q_s+1}, \sigma_{q_s+1})) \leftarrow \mathcal{A}(\mathtt{GUESS} : \mathsf{vk})$;
5. $\mathtt{IF}\ \exists i \neq j, m_i = m_j\ \mathtt{OR}\ \exists i, \mathsf{Verif}(\mathsf{pk}, m_i, \sigma_i) = 0$
   $\mathtt{RETURN}\ 0$
6. $\mathtt{ELSE\ RETURN}\ 1$

Fig. 3. Security games for blind signatures.

In the above games, queries of the adversary are required to be well-formed.

4   The Previous Scheme

In this section, we recall the old scheme in [4] and point out its flaw. From now on, we will work on $\mathbb{F}_2$. The previous scheme is as follows.

KeyGen(k, k', n, n') :
From some integer parameters k, k', n and n', generate:

- $\mathbf{H}$ a trapdoor parity check matrix of size $k \times n$ and its trapdoor $D_{\mathbf{H}}(\cdot)$, only available to the signer $\mathcal{S}$.
- $\mathbf{A}$ a random matrix of size $k \times n'$.
- $\mathbf{B}$ a random matrix of size $k' \times n'$.

Fig. 4. Key generation.

BSProtocol$(w_{\mathbf{x}})$:
1) **Blinding step**
The user $\mathcal{U}$

- generates uniformly at random a vector $\mathbf{x}$ in $\mathbb{F}_2^{n'}$ of weight $w_{\mathbf{x}}$.
- sends $\mu = h(M|\mathbf{Dx}^T) + \mathbf{Ax}^T$ to $\mathcal{S}$.

The signer $\mathcal{S}$ returns $\mathbf{y} = D_{\mathbf{H}}(\mu)$ of weight $w_{\mathbf{y}}$ to the user $\mathcal{U}$. If $D_{\mathbf{H}}(\mu)$ returns $\perp$, then $\mathcal{U}$ can send another request.

2) **Blind Signature step**: $\mathcal{U}$ sends the couple $(\mathbf{Bx}^T, \mathsf{PoK})$, where PoK is a transcript of the proof of knowledge that $\mathcal{U}$ knows a pair of vectors $(\mathbf{x}, \mathbf{y})$ of weight $w_{\mathbf{x}}$ and $w_{\mathbf{y}}$ such that

$$\begin{pmatrix} \mathbf{A} & \mathbf{H} \\ \mathbf{B} & \mathbf{0} \end{pmatrix} \cdot \begin{pmatrix} \mathbf{x}^T \\ \mathbf{y}^T \end{pmatrix} = \begin{pmatrix} h(M|\mathbf{Bx}^T) \\ \mathbf{Bx}^T \end{pmatrix}.$$

The proof of knowledge is obtained through the concatenated ZK Stern protocol of Section 2.3, by taking $\mathbf{Q} = \begin{pmatrix} \mathbf{A} \\ \mathbf{B} \end{pmatrix}$ and $\mathbf{R} = \begin{pmatrix} \mathbf{H} \\ \mathbf{0} \end{pmatrix}$.

Fig. 5. The blind signature protocol.

As mentioned above, with this scheme, there is a flaw in the proof of unforgeability property, that is, we can no longer use an adversary, who can break the soundness of the scheme, to solve the CFS problem or to break the soundness of the underlying zero-knowledge proof. The reason is that after answering queries for the adversary, the simulator still does not know the values $\mathbf{x}$'s. Thus, two problems follow. First, in the rewinding step, "output another random value" does not guarantee that there are no collisions. Note that the queries to the signing oracle are of the form $h(M|\mathbf{Bx}^T) + \mathbf{Ax}^T$, which also contains the term $\mathbf{Ax}^T$. Second, since the simulator does not know the values $\mathbf{x}$'s, which were used by the adversary, there is no way he can accomplish " setting $h(M_{y_j}|B_j)$ to $\mathbf{u}^* - A_{x_j}$." Thus the adversary could not be used to solve the CFS problem.

> **Verification**: Upon receiving the message M and the signature $(\mathbf{B}\mathbf{x}^T, \mathsf{PoK})$, the verifier checks that the proof of knowledge PoK is correct and that the weights $w_\mathbf{x}$ and $w_\mathbf{y}$ of $\mathbf{x}$ and $\mathbf{y}$ are correct.

Fig. 6. Verification protocol.

5   A New Scheme

5.1   The Scheme

In this section, we propose a scheme, which corrects the above flaw. The key generation algorithm remains as in the above scheme, the blind and verification protocols are as follows.

$\mathsf{BSProtocol}(w_\mathbf{x})$:
1) **Blinding step**
The user $\mathcal{U}$

- generates uniformly at random a vector $\mathbf{x}$ in $\mathbb{F}_2^{n'}$ of weight $w_\mathbf{x}$.
- generates $\pi(\mathbf{x})$, a proof of knowledge for $\mathbf{x}$ with respect to $\mathbf{B}\mathbf{x}^T$.
- sends $\mu = h\big(M|\mathbf{B}\mathbf{x}^T|\pi(\mathbf{x})\big) + \mathbf{A}\mathbf{x}^T$ to $\mathcal{S}$.

The signer $\mathcal{S}$ returns $y = D_\mathbf{H}(\mu)$ of weight $w_\mathbf{y}$ to the user $\mathcal{U}$. If $D_\mathbf{H}(\mu)$ returns $\perp$, then $\mathcal{U}$ can send another request.

2) **Blind Signature step**: $\mathcal{U}$ sends the triple $(\mathbf{B}\mathbf{x}^T, \pi(\mathbf{x}), \mathsf{PoK})$, where PoK is a transcript of the proof of knowledge that $\mathcal{U}$ knows a pair of vectors $(\mathbf{x}, \mathbf{y})$ of weight $w_\mathbf{x}$ and $w_\mathbf{y}$ such that

$$\begin{pmatrix} \mathbf{A} & \mathbf{H} \\ \mathbf{B} & \mathbf{0} \end{pmatrix} \cdot \begin{pmatrix} \mathbf{x}^T \\ \mathbf{y}^T \end{pmatrix} = \begin{pmatrix} h\big(M|\mathbf{B}\mathbf{x}^T|\pi(\mathbf{x})\big) \\ \mathbf{B}\mathbf{x}^T \end{pmatrix}.$$

The proof of knowledge is obtained through the concatenated ZK Stern protocol by taking $\mathbf{Q} = \begin{pmatrix} \mathbf{A} \\ \mathbf{B} \end{pmatrix}$ and $\mathbf{R} = \begin{pmatrix} \mathbf{H} \\ \mathbf{0} \end{pmatrix}$.

Fig. 7. The corrected blind signature protocol.

> **Verification**: Upon receiving the message M and the signature $(\mathbf{B}\mathbf{x}^T, \pi(\mathbf{x}), \mathsf{PoK})$, the verifier checks that the proofs of knowledge $\pi(\mathbf{x}), \mathsf{PoK}$ are correct and that the weights $w_\mathbf{x}$ and $w_\mathbf{y}$ of $\mathbf{x}$ and $\mathbf{y}$ are correct.

Fig. 8. The corrected verification protocol.

In this scheme, we add a proof of knowledge of $\mathbf{x}$ in the hash queries, *i.e.*, a hash query consists of a message M, the value $\mathbf{Bx}^T$, and a Stern-like proof of knowledge $\pi(\mathbf{x})$ of $\mathbf{x}$ with respect to $\mathbf{Bx}^T$. In order to obtain this proof, one has to query the random oracle three times to get the values of commitments. From these queries, the one in control of the random oracle would certainly know the value $\mathbf{x}$. This argument guarantees that in our proof, the simulator would know the values $\mathbf{x}$'s and with this knowledge, he could efficiently manipulate the random oracle to avoid collisions. Note that $\pi(\mathbf{x})$ only needs to have one round. The verification protocol is considered to be successful if both $\pi(\mathbf{x})$ and PoK are valid.

5.2   Unforgeability

Theorem 2. *If there exists an adversary against the soundness of the blind signature scheme, then there exists an adversary for either the CFS Problem, the syndrome decoding problem, or the soundness of the underlying zero-knowledge proof.*

Proof. If an adversary $\mathcal{A}$ can win the game of unforgeability of the blind signature, then he can produce $N+1$ blind signatures with N requests to the blind oracle. To exploit this adversary, we build a simulator in the following way. We first receive the matrix $\mathbf{H}$ and a hash function h from the challenge oracle for CFS problem and generate normally the other parameter of our blind signature. The hash and signing queries are treated as follows.

- Receiving signing queries, on string $\mathbf{c}_i$, we forward it to the CFS oracle, and receive y_i such that $\mathbf{Hy}_i^T = \mathbf{c}_i^T$.
- Receiving hash queries, the simulator answers with a random value, and stores it to answer in the same way to similar queries.

After at most N signing queries and $\mathfrak{n}$ random oracle queries, the adversary sends us $N+1$ signatures σ_j on messages M_j, by sending us values B_j, π_j, and zero-knowledge proofs PoK_j, that he knows $\mathbf{x}_j, \mathbf{y}_j$ such that $B_j = \mathbf{Bx}_j^T, \pi_j = \pi(\mathbf{x}_j)$, and $\mathbf{Hy}_j^T = h(M_j|B_j|\pi_j) - \mathbf{Ax}_j^T$. (We are working on $\mathbb{F}_2$ so $\mathbf{Ax}_j^T = -\mathbf{Ax}_j^T$.) As this is a valid forgery against the blind signature scheme, then all the $N+1$ signatures are valid. This means that either the adversary manages to break the soundness of one of the proofs, or by using the random oracle, the simulator manages to extract the values $\mathbf{x}_j, \mathbf{y}_j$.

If two values $\mathbf{y}_{j_1}$ and $\mathbf{y}_{j_2}$ are equal, then the adversary has managed to find a collision on $h(M_{j_b}|B_{j_b}|\pi_{j_b}) - A_{j_b}$, where $A_{j_b} = \mathbf{Ax}_{j_b}^T$. In this case, we simply rewind to the furthest random oracle query on $M_{j_b}|B_{j_b}|\pi_{j_b}$ and output another random value such that there is no longer a collision neither with the query corresponding to j_b, nor with any queries done before to the random oracle. Note that the queries contain $\pi(\mathbf{x})$, a proof of knowledge of $\mathbf{x}$ with respect to $\mathbf{Bx}^T$. In order to obtain these proofs, the adversary has to query the random oracle on the values of commitments. In this way, the simulator always knows the pairs $(\mathbf{x}, \mathbf{Bx}^T)$, as long as $\mathcal{A}$ would like to generate $\pi(\mathbf{x})$. Now, the forking lemma ensures us that the adversary's advantages is approximately the same, after k rewinding where k is upper-bounded by $\min(N, \mathfrak{n})$.

After this, we are sure that all the $N + 1$ values $\mathbf{y}_j$'s are different, so there exists at least one $\mathbf{y}_j$ that does not come from the challenge oracle. Rewinding one last time, and setting $h(M_{y_j}|B_j|\pi_j)$ to $\mathbf{u}^{*T} - \mathbf{A}\mathbf{x}_j^T$ (this can always be done since the simulator knows the value $\mathbf{x}_j$), and invoking the forking lemmas, allows to recover an $\mathbf{y}_j$ such that $\mathbf{H}\mathbf{y}_j^T = h(\mathbf{u}^*)$ and so it allows to solve the CFS challenge.   □

5.3   Blindness

Theorem 3. *If there exists an adversary against the blindness of the blind signature scheme, then there exists an adversary against the zero-knowledge property of the Stern protocol or the decisional syndrome decoding problem.*

Proof. If an adversary $\mathcal{A}$ can win the game of blindness of the blind signature scheme, then he can break the decisional syndrome decoding problem. To exploit this adversary, we build a simulator in the following way. We first receive a decisional syndrome decoding instance $\mathbf{C}, \mathbf{s}$ and have to guess whether there exists a small $\mathbf{x}$ such that $\mathbf{C} \cdot \mathbf{x}^T = \mathbf{s}^T$. The simulator splits the matrix $\mathbf{C}$ into $\mathbf{A}$ and $\mathbf{B}$ of size $k \times n'$ and $k' \times n'$, respectively (as in the scheme), generates a matrix $\mathbf{H}$ honestly and publishes them as the public keys of the scheme, and gives $\mathbf{H}$'s trapdoor to the adversary. The adversary then sends two messages M_0 and M_1 to the simulator. The simulator picks a random bit $b \leftarrow \{0, 1\}$, and proceeds to send the requests on M_b and M_{1-b}, and then outputs the signature on M_0.

With advantage ϵ, the adversary guesses whether $b = 0$ or not. Next, the simulator proceeds to a sequence of games.

Game G_1. In this game, the simulator proceeds honestly, however, instead of outputting the real $\pi(\mathbf{x}_0)$ and PoK_0, he outputs simulated proofs π_0 and Π_0. At this step, the adversary's view is $\mathbf{B}\mathbf{x}_0^T, \pi_0$, and $h(M_b|\mathbf{B}\mathbf{x}_b^T|\pi^*) + \mathbf{A}\mathbf{x}_b^T$, where $\pi^* \in \{\pi_0, \pi(\mathbf{x}_b)\}$. We can assume that $\mathbf{A}\mathbf{x}_1^T + h(M_1|\mathbf{B}\mathbf{x}_1^T|\pi(\mathbf{x}_1)) \neq h(M_0|\mathbf{B}\mathbf{x}_0^T|\pi(\mathbf{x}_0)) + \mathbf{A}\mathbf{x}_0^T$. (Controlling the random oracle allows to make sure of that, anyway it happens with overwhelming probability.)

Game G_2. In this game, the simulator makes the following change. He splits $\mathbf{s}$ into $\mathbf{s}_1, \mathbf{s}_2$, sets the value of $\mathbf{A}\mathbf{x}_0^T$ to be equal to $\mathbf{s}_1^T$, and the value of $\mathbf{B}\mathbf{x}_0^T$ to be equal to $\mathbf{s}_2^T$.

We analyze the answer of the adversary as follows. If the answer to the challenge was yes, we are still in the previous game G_1. On the contrary, if it was no, it leads us to the last game G_2, where the vector $\mathbf{s}$ does not come from the SD distribution. The last game G_2 yields a completely simulated answer, with random public values, so the adversary has no advantage against the blindness in G_2. The difference between G_2 and G_1 is the decisional syndrome decoding problem, while the zero-knowledge property differentiates G_1 from the real game. Hence $\epsilon \leq \mathsf{Adv}_{ZK} + \mathsf{Adv}_{DSD}$. Therefore, there is either an adversary against the DSD problem or the zero-knowledge property of the Stern protocol.   □

5.4   Parameters

Overall, the best practical attacks against forgery is the attack against the invertible trapdoor function $D_{\mathbf{H}}(\cdot)$, and the best practical attack for blindness is retrieving a small weight vector $\mathbf{x}$ of weight $w_{\mathbf{x}}$ from the syndrome $\mathbf{Bx}^T$, for a random matrix $\mathbf{B}$. Hence, we choose parameters according to these constraints. The size of the public key is $P = kn + (k + k')n'$. The size of the signature is the total size of $\mathbf{Bx}^T, \pi(\mathbf{x})$, and PoK, which is

$$S = k' + n'(\log n' + 1) + \ell \cdot \left(n(\log n + 1) + n'(\log n' + 1) + 5\lambda\right),$$

where ℓ satisfies $(2/3)^\ell = 2^{-\lambda}$ for λ the security parameter.

We now give example of parameters for our scheme, considering parameters for which a word of weight $w_{\mathbf{x}}$ is unique with very strong probability:

We consider the parallel CFS signature scheme with parameters $n = 2^{18}$, $w_{\mathbf{y}} = 9$ and $k = 162$, $n' = 6000$, $k' = 300$ and $w_{\mathbf{x}} = 30$. For that case, the security of parallel CFS is 2^{82} and 2^{91} for the cost of recovering a unique (with strong probability) $\mathbf{x}$ of weight 30 from its syndromes by matrices $\mathbf{A}$ and $\mathbf{B}$. We choose $\lambda = 80$ and $\ell = 137$ so the size of public key is $P = 5.65$ MB, the size of signature is $S = 86.7$ MB.

6   Conclusion

We have proposed a new blind signature scheme to repair the one proposed by Blazy et al. (IEEE 2017). In general, the size of public key and signature differ only slightly from that of the previous scheme. Only the signature size increases a bit due to the addition of the proof of knowledge of the randomness.

The blinding step of our scheme makes use of the trapdoor function $D_{\mathbf{H}}(\cdot)$ of a CFS signature scheme. It might be tempting to try another primitives such that Durandal [2] or WAVE [7]. We leave these speculations for future work.

Acknowledgements. The authors would like to thank anonymous reviewers for their helpful comments on the paper and also Damien Stehlé, Shweta Agrawal, and Anshu Yadav for pointing out the mistake in the original construction.

References

1. Alamélou, Q., Blazy, O., Cauchie, S., Gaborit, P.: A code-based group signature scheme. Des. Codes Cryptogr. **82**(1-2), 469–493 (2017)
2. Aragon, N., Blazy, O., Gaborit, P., Hauteville, A., Zémor, G.: Durandal: a rank metric based signature scheme. In: Ishai, Y., Rijmen, V. (eds.) EUROCRYPT 2019. LNCS, vol. 11478, pp. 728–758. Springer, Cham (2019). https://doi.org/10.1007/978-3-030-17659-4_25
3. Applebaum, B., Ishai, Y., Kushilevitz, E.: Cryptography with constant input locality. In: Menezes, A. (ed.) CRYPTO 2007. LNCS, vol. 4622, pp. 92–110. Springer, Heidelberg (2007). https://doi.org/10.1007/978-3-540-74143-5_6

4. Blazy, O., Gaborit, P., Schrek, J., Sendrier, N.: A code-based blind signature. In: IEEE International Symposium on Information Theory (2017)
5. Chaum, D.: Blind signatures for untraceable payments. In: Chaum, D., Rivest, R.L., Sherman, A.T. (eds.) Advances in Cryptology, pp. 199–203. Springer, Boston, MA (1983). https://doi.org/10.1007/978-1-4757-0602-4_18
6. Courtois, N.T., Finiasz, M., Sendrier, N.: How to achieve a McEliece-based digital signature scheme. In: Boyd, C. (ed.) ASIACRYPT 2001. LNCS, vol. 2248, pp. 157–174. Springer, Heidelberg (2001). https://doi.org/10.1007/3-540-45682-1_10
7. Debris-Alazard, T., Sendrier, N., Tillich, J.-P.: Wave: a new family of trapdoor one-way preimage sampleable functions based on codes. In: Galbraith, S.D., Moriai, S. (eds.) ASIACRYPT 2019. LNCS, vol. 11921, pp. 21–51. Springer, Cham (2019). https://doi.org/10.1007/978-3-030-34578-5_2
8. Faugère, J.-C., Gauthier, V., Otmani, A., Perret, L., Tillich, J.-P.: A distinguisher for high rate McEliece cryptosystems. In: ITW 2011, pp. 282–286. Paraty, Brazil (October 2011)
9. Fiat, A., Shamir, A.: How to prove yourself: practical solutions to identification and signature problems. In: Odlyzko, A.M. (ed.) CRYPTO 1986. LNCS, vol. 263, pp. 186–194. Springer, Heidelberg (1987). https://doi.org/10.1007/3-540-47721-7_12
10. Finiasz, M.: Parallel-CFS: strengthening the CFS McEliece-based signature scheme. In: Biryukov, A., Gong, G., Stinson, D.R. (eds.) Selected Areas in Cryptography. SAC 2010. LNCS, vol. 6544, pp. 159–170. Springer, Berlin, Heidelberg (2011). https://doi.org/10.1007/978-3-642-19574-7_11
11. Fischlin, M.: Round-optimal composable blind signatures in the common reference string model. In: Dwork, C. (ed.) CRYPTO 2006. LNCS, vol. 4117, pp. 60–77. Springer, Heidelberg (2006). https://doi.org/10.1007/11818175_4
12. Garg, S., Gupta, D.: Efficient round optimal blind signatures. In: Nguyen, P.Q., Oswald, E. (eds.) EUROCRYPT 2014. LNCS, vol. 8441, pp. 477–495. Springer, Heidelberg (2014). https://doi.org/10.1007/978-3-642-55220-5_27
13. Hauck, E., Kiltz, E., Loss, J., Nguyen, N.K.: Lattice-based blind signatures, revisited. In: Micciancio, D., Ristenpart, T. (eds.) CRYPTO 2020. LNCS, vol. 12171, pp. 500–529. Springer, Cham (2020). https://doi.org/10.1007/978-3-030-56880-1_18
14. Landais, G., Sendrier, N.: Implementing CFS. In: Galbraith, S., Nandi, M. (eds.) INDOCRYPT 2012. LNCS, vol. 7668, pp. 474–488. Springer, Heidelberg (2012). https://doi.org/10.1007/978-3-642-34931-7_27
15. Okamoto, T.: Provably secure and practical identification schemes and corresponding signature schemes. In: Brickell, E.F. (ed.) CRYPTO 1992. LNCS, vol. 740, pp. 31–53. Springer, Heidelberg (1993). https://doi.org/10.1007/3-540-48071-4_3
16. Overbeck, R.: A step towards QC blind signatures. In: Cryptology ePrint Archive: Report 2009/102. https://eprint.iacr.org/2009/102.pdf
17. Pointcheval, D., Stern, J.: New blind signatures equivalent to factorization. In: Proceedings of the 4th CCCS, pp. 92–99. ACM Press, New York (1997)
18. Pointcheval, D., Stern, J.: Security arguments for digital signatures and blind signatures. J. Cryptol. 13(3), 361–396 (2000)
19. Rückert, M.: Lattice-based blind signatures. In: Abe, M. (ed.) ASIACRYPT 2010. LNCS, vol. 6477, pp. 413–430. Springer, Heidelberg (2010). https://doi.org/10.1007/978-3-642-17373-8_24
20. Stern, J.: A new paradigm for public key identification. IEEE Trans. Inf. Theory 42(6), 1757–1768 (1996)
21. Stern, J.: A new identification scheme based on syndrome decoding. In: Stinson, D.R. (ed.) CRYPTO 1993. LNCS, vol. 773, pp. 13–21. Springer, Heidelberg (1994). https://doi.org/10.1007/3-540-48329-2_2

Performance Bounds for QC-MDPC Codes Decoders

Marco Baldi[1], Alessandro Barenghi[2], Franco Chiaraluce[1], Gerardo Pelosi[2], and Paolo Santini[1(✉)]

[1] Università Politecnica delle Marche, Ancona, Italy
{m.baldi,f.chiaraluce,p.santini}@univpm.it
[2] Politecnico di Milano, Milano, Italy
{alessandro.barenghi,gerardo.pelosi}@polimi.it

Abstract. Quasi-Cyclic Moderate-Density Parity-Check (QC-MDPC) codes are receiving increasing attention for their advantages in the context of post-quantum asymmetric cryptography based on codes. However, a fundamentally open question concerns modeling the performance of their decoders in the region of a low decoding failure rate (DFR). We provide two approaches for bounding the performance of these decoders, and study their asymptotic behavior. We first consider the well-known Maximum Likelihood (ML) decoder, which achieves optimal performance and thus provides a lower bound on the performance of any sub-optimal decoder. We provide lower and upper bounds on the performance of ML decoding of QC-MDPC codes and show that the DFR of the ML decoder decays polynomially in the QC-MDPC code length when all other parameters are fixed. Secondly, we analyze some hard to decode error patterns for Bit-Flipping (BF) decoding algorithms, from which we derive some lower bounds on the DFR of BF decoders applied to QC-MDPC codes.

Keywords: QC-MDPC codes · Decoding failure rate · Bit-Flipping decoder · Maximum likelihood decoder · Error floor · Post-quantum cryptography · Code-based cryptography

1 Introduction

Code-based public-key cryptography is deemed as one of the most consolidated and promising areas in post-quantum cryptography. As the most remarkable example, we can mention the Classic McEliece scheme [2], which currently appears as a finalist in the NIST post-quantum standardization process [1,26]. This scheme essentially consists of a highly optimized version of the original proposal by Robert McEliece [25] and, in particular, employs the same family of error correcting codes (namely, binary Goppa codes). Despite more than 40 years of cryptanalysis, the improvements in known attacks against the original McEliece scheme, which are substantially based on Information Set Decoding (ISD) algorithms, have been very limited (see [5] for a review of such algorithms, and [10,12] for the state of the art). However, this robustness is somehow paid

© Springer Nature Switzerland AG 2022
A. Wachter-Zeh et al. (Eds.): CBCrypto 2021, LNCS 13150, pp. 95–122, 2022.
https://doi.org/10.1007/978-3-030-98365-9_6

with very large public keys, a feature that has historically represented Achille's heel of code-based cryptography and, ultimately, has hindered its spreading in modern applications.

To overcome this issue, researchers have thoroughly investigated the possibility of replacing Goppa codes with other error correcting codes, and/or that of adding some geometrical structure to the employed codes, which may enable a more compact code representation. However, the majority of such attempts were unsuccessful, either because of algebraic attacks (such as [14,37]), structural attacks (such as [3,13,24]), or a combination of them [19]. While algebraic code structures proved more difficult to hide and have led to unbroken instances with moderate advantages in terms of public key size [9,23], more important reductions in the key size can be achieved by resorting to random-based structured codes like Quasi-Cyclic Moderate-Density Parity-Check (QC-MDPC) codes [4,29], which derive from the well-known family of Quasi-Cyclic Low-Density Parity-Check (QC-LDPC) codes [6]. However, low-complexity decoding of QC-MDPC codes, as well as QC-LDPC codes, is performed through iterative algorithms derived from Gallager's Bit Flipping (BF) decoder [20] and, differently from bounded distance decoders used for algebraic codes like Goppa codes, these algorithms are characterized by a non-null Decoding Failure Rate (DFR). This implies that an adversary performing a Chosen-Ciphertext Attack (CCA) can gather information about the secret key by exploiting decryption failures [18,21,30]. Formally, this translates into the fact that a non-zero DFR may prevent the cryptosystem from achieving Indistinguishability under Adaptively Chosen Ciphertext Attack (IND-CCA2), that is, resistance against active attackers, which is fundamental in many scenarios. To overcome this issue, it is required to guarantee that the decoding algorithm has a provably low DFR, namely, not higher than $2^{-\lambda}$, where λ is the target security level in bits [22]. Such small values of DFR are impossible to assess directly through numerical simulations; thus, finding theoretical models for the DFR of decoders for QC-MDPC codes is of paramount importance.

Related Works. For a single-iteration BF decoder, the available analyses establish both the error correction capability [32,38] and a provable, code-specific upper bound on the DFR [31]. However, for more than one decoder iteration, these models require some assumptions that result in a loose modeling of the decoder performance. Multiple iterations have been conservatively analyzed in [7], for a decoder that however processes the bits in a sequential manner and, consequently, is not efficient in practice. Following a completely different strategy, authors in [35] study the dependence of the DFR on the code length, and propose to extrapolate such a function in the region of low DFR values, based on its trend estimated through numerical simulations for smaller DFR values. Such an extrapolated performance is then used to adjust the code and decoder parameters [16,17], as well as to design parameter sets for the BIKE cryptosystem [4]. A theoretical justification of this approach is provided in [34], where the authors claim that the logarithm of the DFR is a concave function of the code length, up to the point where the DFR is not larger than $2^{-\lambda}$. Under this assumption, extrapolation with an exponential decay in the code length yields a conservative DFR estimate. This is motivated in [34] through the assumption that, when the DFR is extremely low, the only relevant failure phenomena in a BF decoder are those due to input sequences for which the closest codeword is different from the transmitted one. Notice that, if such an assumption is true, then the BF decoder must approach the optimal (Maximum Likelihood (ML)) decoder in the region of low DFR.

Our Contribution. We study the performance of decoders for QC-MDPC codes in the setting with a fixed number of errors. We start by analyzing the DFR of the optimum decoding strategy, corresponding to a complete ML decoder which additionally exploits the knowledge on the number of introduced errors. We show that, for some families of QC-MDPC codes (like those employed in BIKE), this decoder is characterized by a non-zero DFR that decays polynomially in the code length (assuming all the other parameters as fixed). Studying the performance of ML decoding allows us to obtain a lower bound on the DFR of any sub-optimal decoder. In particular, through our analysis, we are able to formally and rigorously prove the existence of the *error floor* region, for the considered codes, as a function of the code length. The error floor is a well-known phenomenon when the codes are used in communication systems, for example affected by thermal noise, but its dependence on the code length has been rarely investigated in previous literature [28, 34].

More precisely, we consider BF decoders for QC-MDPC codes and show how to identify some error patterns that, with high probability, cannot be corrected. By doing this, we are able to compute a lower bound on the DFR of such decoders. With our results, we are able to provide evidence in contrast with the claims in [34], in particular showing that: i) a BF decoder is extremely far from being optimal, and ii) the most likely failure events are not those due to near-codewords. It must be noted that, in an independent and very recent work [41], Vasseur has come to similar conclusions with a thorough analysis of near-codewords and their impact on the DFR. We remark that our results do not directly imply that the parameters proposed in [4, 17, 34] do not achieve the claimed DFR. However, they suggest that finding exact models for the performance of a BF decoder still requires further investigations, especially in the regime, here of interest, of extremely low DFR.

The paper is organized as follows. In Sect. 2 we establish the notation used throughout the paper, and we provide basic concepts about coding theory and QC-MDPC codes. In Sect. 3 we analyze the ML decoder, and we employ the obtained results to prove the existence of the floor for specific families of QC-MDPC codes. In Sect. 4 we take into account BF decoding, and we describe how to pick hard to decode errors and how to use such vectors to find a lower bound on the DFR. Finally, in Sect. 5 we draw some concluding remarks.

2   Notation and Background

We use bold uppercase letters to denote matrices, and bold lowercase letters to denote vectors. Given a matrix $\mathbf{A}$, we use $\mathbf{A}_{i,:}$ (resp, $\mathbf{A}_{:,i}$) to denote its i-th row (resp. column), while $a_{i,j}$ refers to its entry in the i-th row and j-th column. For a vector $\mathbf{a}$, we use a_i to refer to its i-th component. The null vector of length n is indicated as $\mathbf{0}_n$. The Hamming weight of vector $\mathbf{a}$ is denoted as $\mathrm{wt}(\mathbf{a})$, while $\mathrm{Supp}(\mathbf{a})$ refers to its support, that is, the set of indexes pointing at non-null entries. Let $\mathbb{F}_2$ denote the binary finite field. For two vectors $\mathbf{a}$ and $\mathbf{b}$ with equal length, defined over $\mathbb{F}_2$, we denote as $\langle \mathbf{a} ; \mathbf{b} \rangle$ their integer inner product, that is, their inner product after lifting their entries from $\mathbb{F}_2$ to the ring of integers $\mathbb{Z}$.

For a set A, the expression $a \xleftarrow{\$} A$ means that a is uniformly picked among the elements of A; the cardinality of the set is denoted as $|A|$. We use $B_{n,w} \subset \mathbb{F}_2^n$ to denote the Hamming sphere with radius w, that is, the set of length-n vectors with Hamming weight w.

2.1   Error Correcting Codes

In the following we focus on linear block codes over $\mathbb{F}_2$.

Definition 1. *A linear code $\mathscr{C}$ of length n, dimension k and redundancy $r = n - k$ over $\mathbb{F}_2$ is a k-dimensional linear subspace of $\mathbb{F}_2^n$. We say that $\mathbf{G} \in \mathbb{F}_2^{k \times n}$ is a generator matrix for $\mathscr{C}$ if it is a basis of $\mathscr{C}$; a matrix $\mathbf{H} \in \mathbb{F}_2^{r \times n}$ is said to be a parity-check matrix for $\mathscr{C}$ if it is a basis of its null space.*

A crucial property of a linear code is that the sum of any number of codewords yields another codeword. Codes are normally endowed with a distance metric, that is, a function able to measure the distance between pairs of codewords; in this paper we only consider the Hamming metric, defined next.

Definition 2. *The Hamming distance in the vector space $\mathbb{F}_2^n$ is defined as the function* $\mathrm{dist} : \mathbb{F}_2^n \times \mathbb{F}_2^n \to \mathbb{N}$ *such that*

$$\mathrm{dist}(\mathbf{a}, \mathbf{b}) = |\mathrm{Supp}(\mathbf{a} + \mathbf{b})| = \mathrm{wt}(\mathbf{a} + \mathbf{b}).$$

We finally recall the concepts of *weight distribution* and *minimum distance*.

Definition 3. *For a linear code $\mathscr{C} \subseteq \mathbb{F}_2^n$ and $w \in [0; n]$, we denote with A_w the number of codewords whose weight is w. Then, the weight distribution of $\mathscr{C}$ corresponds to the collection of all values A_w. The minimum distance of $\mathscr{C}$ is defined as the minimum $w > 0$ such that $A_w > 0$ or, equivalently, as*

$$d = \min\left\{\mathrm{dist}(\mathbf{c}, \mathbf{c}') \mid \mathbf{c}, \mathbf{c}' \in \mathscr{C}, \ \mathbf{c} \neq \mathbf{c}'\right\} = \min\{\mathrm{wt}(\mathbf{c}) \mid \mathbf{c} \in \mathscr{C} \setminus \mathbf{0}_n\}.$$

Arguably, the most important application of linear codes is that of error correction over noisy channels; this is accomplished through decoding algorithms, i.e., techniques that, within certain limits, can identify the channel action on a received sequence and, consequently, reconstruct the transmitted codeword. In this work, we focus on the use of error correcting codes in the context of the McEliece cryptosystem. In such a setting, the message to be transmitted is first encoded as a codeword and then an *error vector* **e** of fixed weight t is added to it. To this end, we introduce the *McEliece channel*, whose action is described as

$$\mathbf{c} \mapsto \mathbf{x} = \mathbf{c} + \mathbf{e}, \quad \mathbf{c} \in \mathbb{F}_2^n, \quad \mathbf{e} \xleftarrow{\$} B_{n,t},$$

where **c** is the input sequence and **e** is the error introduced by the channel.

To provide a rigorous classification of decoders, we consider the following formal definition, which has been made specific to the McEliece channel.

Definition 4. *Let $\mathscr{C} \subseteq \mathbb{F}_2^n$ be a linear code of length n. We say that an algorithm* Dec $: \mathbb{F}_2^n \to \mathbb{F}_2^n$ *has DFR ϵ for $\mathscr{C}$, in the McEliece channel with parameter t, if*

$$\Pr\left[\mathsf{Dec}(\mathbf{c} + \mathbf{e}) \neq \mathbf{c} \;\middle|\; \mathbf{c} \xleftarrow{\$} \mathscr{C}, \; \mathbf{e} \xleftarrow{\$} B_{n,t}\right] = \epsilon.$$

2.2 QC-MDPC Codes

Let us recall the definition of MDPC codes, which were first introduced in [27] for the context of communications but, later on, received interest for the use in public-key cryptosystems [29].

Definition 5. *Let $\mathbf{H} \in \mathbb{F}_2^{r \times n}$ such that all of its rows have weight $w = O(\sqrt{n})$; then, we say that the code having $\mathbf{H}$ as parity-check matrix is an MDPC code.*

Namely, MDPC codes are analogous to LDPC codes, with the only difference that their parity-check matrices are denser than those of typical LDPC codes.

In particular, when used in cryptography, these codes are usually endowed with the QC structure, that is, the matrix **H** is formed by circulant blocks of size p. Note that, for a circulant matrix, all rows and columns have the same weight; thus, with some abuse of notation, we will use the term "weight of a circulant matrix" to refer to the weight of any of its rows/columns. From now on, we will focus on a particular class of QC-MDPC codes, which we formally define as follows.

Definition 6. *Let $\mathbf{H} = [\mathbf{H}_0, \cdots, \mathbf{H}_{n_0-1}]$, with $\mathbf{H}_i$ being a circulant matrix of size p and weight $v = O(\sqrt{p/n_0})$. Then, we say that the code $\mathscr{C}$ admitting $\mathbf{H}$ as parity-check matrix is an n_0-QC-MDPC code. Furthermore, we denote with $\mathcal{QC}\text{-}\mathcal{MDPC}(n_0, p, v)$ the collection of all such codes.*

Note that an n_0-QC-MDPC code has length $n = n_0 p$, dimension $k = (n_0 - 1)p$ and redundancy $r = p$. A parity-check matrix as in Definition 6 has all columns with weight v, while all rows have weight $w = n_0 v = O(\sqrt{n})$. In a cryptosystem, a user randomly and uniformly picks a code from $\mathcal{QC}\text{-}\mathcal{MDPC}(n_0, p, v)$, and uses its parity-check matrix as the secret key.

3 Maximum-Likelihood Decoding

In this section we analyze the optimal decoding strategy of QC-MDPC codes exploiting ML, and characterize its performance over the McEliece channel. Such a technique works by first testing the distance between each codeword and the received sequence, and then by outputting the codeword that minimizes such a distance. When there is more than one codeword at the same minimum distance from the received sequence, then the ML decoder can apply one of the following two strategies:

- *Complete* ML decoding: the decoder randomly outputs one of the codewords at minimum distance from the received sequence.
- *Incomplete* ML decoding: the decoder halts and reports a decoding failure.

In this paper we consider a complete ML decoder. We observe that the results we obtain can easily be adapted to the case of an incomplete ML decoder and, in general, no big difference exists between the two behaviors from a practical standpoint. However, since the complete ML decoder always returns a codeword, it is clear that its DFR is lower than that of the incomplete counterpart. Indeed, the two decoders behave differently only when there is more than one codeword at the same distance from the received sequence. In such a situation, the incomplete decoder will not try to decode (hence, according to Definition 4, it will fail), while with some non null probability the complete version will return the correct codeword.

Taking into account the fact that, in our case, there are exactly t errors affecting each transmitted codeword, we can modify the standard definition of complete ML decoding as follows.

Definition 7. *Let $\mathscr{C}$ be a linear code over $\mathbb{F}_2$ with length n. The complete ML-decoder is the algorithm* ML $: \mathbb{F}_2^n \to \mathscr{C}$ *that, on input $\mathbf{x} \in \mathbb{F}_2^n$, returns $\mathbf{c}' \overset{\$}{\leftarrow} \mathscr{C}^{(\mathbf{x})}$, where*

$$\mathscr{C}^{(\mathbf{x})} = \left\{ \mathbf{c} \in \mathscr{C} \text{ s.t. } \mathrm{dist}(\mathbf{x}, \mathbf{c}) = t \right\},$$

that is, $\mathscr{C}^{(\mathbf{x})}$ is the set of all the codewords of $\mathscr{C}$ which are exactly t away from $\mathbf{x}$ under Hamming distance.

Note that, when $\mathscr{C}^{(\mathbf{x})}$ contains only one codeword, obviously that codeword is the decoder output (so, no randomness is involved). We point out that the decoder we have defined above corresponds to the best decoder (in terms of DFR) one can dispose of, in the McEliece channel. Indeed, the decoder i) exploits knowledge on the number of errors, and ii) always returns a codeword. Because of these reasons, the study of its performances is meaningful since it allows us to derive the minimum DFR that can be reached. Complete ML decoding can also be performed when the decoder input is the syndrome of the received sequence; formally, we define such a procedure as follows.

Definition 8. *Let $\mathscr{C}$ be a linear code over $\mathbb{F}_2$ with length n and parity-check matrix* **H**. *We define the* ML *syndrome decoder as the algorithm* MLS: $\mathbb{F}_2^n \rightarrow \mathbb{F}_2^n$ *that, on input* $\mathbf{x} \in \mathbb{F}_2^n$, *returns* $\mathbf{x} + \mathbf{e}'$, *where* $\mathbf{e}' \stackrel{\$}{\leftarrow} \mathcal{S}_{\mathbf{H}}^{(\mathbf{x})}$ *and*

$$\mathcal{S}_{\mathbf{H}}^{(\mathbf{x})} = \{\mathbf{e} \in B_{n,t} \text{ s.t. } \mathbf{e}\mathbf{H}^\top = \mathbf{x}\mathbf{H}^\top\}.$$

Notice that the ML and MLS decoders are, in principle, different from each other. Indeed, the ML decoder always returns a codeword, while the MLS decoder may return a vector that does not belong to the code. Yet, in the following theorem we prove that the DFR of these algorithms coincide, and furthermore we provide explicit bounds for such a failure probability.

Theorem 1. *Let $\mathscr{C} \subseteq \mathbb{F}_2^n$ be a linear code of length n, dimension k and minimum distance d, and consider the transmission over the McEliece channel with parameter t. Then, the ML and MLS decoding algorithms have the same DFR, denoted as ϵ_{ML}, which is equal to*

$$\epsilon_{\mathsf{ML}} = 1 - \frac{1}{\binom{n}{t}} \sum_{\mathbf{e} \in B_{n,t}} \frac{1}{\left| \mathscr{C}^{(\mathbf{e})} \right|}.$$

Furthermore, it holds that $\epsilon_{\mathsf{ML}}^{(L)} < \epsilon_{\mathsf{ML}} < \epsilon_{\mathsf{ML}}^{(U)}$, where

$$\epsilon_{\mathsf{ML}}^{(U)} = \frac{1}{2\binom{n}{t}} \sum_{\substack{w \in [d; 2t] \\ w \text{ even}}} A_w \binom{w}{w/2} \binom{n-w}{t-w/2},$$

$$\epsilon_{\mathsf{ML}}^{(L)} = \begin{cases} 0 & \text{if } A_w = 0 \text{ for all even } w \in [d; 2t], \\ \max_{\substack{w \in [d; 2t] \\ w \text{ even} \\ A_w > 0}} \left\{ \frac{\binom{w}{w/2}\binom{n-w}{t-w/2}}{2\binom{n}{t}} \right\} & \text{otherwise.} \end{cases}$$

Proof. See Appendix A.

It is clear that the computational complexity of both ML and MLS decoders is intractable unless the code or the channel have trivial parameters (i.e., very low values of k and/or t). Indeed, a straightforward implementation of the ML decoder runs in time $O(2^{Rn})$ (since all codewords must be tested), being $R = k/n$ the code rate, while the MLS decoder takes time $O(n^t)$ (since it tests all vectors in $B_{n,t}$, of size $\binom{n}{t} = O(n^t)$). Furthermore, we recall that solving the decoding problem for a generic random linear code was proven to be NP-complete [11], as well as finding its minimum distance [40]. It is thus rather unlikely that efficient implementations of ML decoders are found. For this reason, one normally relies on sub-optimal decoding strategies. Hence, any such practical decoder is going to have a DFR higher than that of the ML decoder.

3.1   ML Decoders for QC-MDPC Codes

When QC-MDPC codes are employed in public-key cryptosystems [4,8,27], we have that both the secret and the public keys are representation of the same code $\mathscr{C}$, drawn at random from $QC\text{-}MDPC(n_0, p, v)$. In particular, the secret key corresponds to a sparse parity-check matrix, while the public key is either a dense generator or a dense parity-check matrix. Furthermore, we have that n_0 is normally chosen as a small integer, namely, $n_0 \in \{2, 3, 4\}$. Because of the QC structure, we can derive some common properties for these codes, as stated in the following proposition.

Proposition 1. *Let $\mathscr{C}$ be picked at random from $QC\text{-}MDPC(n_0, p, v)$. Then, the following properties hold:*

i) the minimum distance of $\mathscr{C}$ is not greater than $2v$;
ii) we have $A_{2v} \geq p\binom{n_0}{2}$.

Proof. See Appendix B.

When employed in a public-key cryptosystem, the parameters of a QC-MDPC code must satisfy some constrains in order to guarantee the desired security level λ. As it is well known, the best attacks against these schemes exploit Information Set Decoding (ISD) algorithms, which are techniques originally conceived for decoding arbitrary codes, when no efficient decoding algorithm is available. Given a code with length n and dimension k, an ISD algorithm can be used to decode an error vector of weight ω with a computational complexity that is well approximated [39] as

$$C_{\mathsf{ISD}}(n, k, \omega) \approx 2^{-\omega \log_2(1-k/n)}.$$

Note that the above complexity also corresponds to that of finding a specific codeword of weight ω in a code with the same parameters. In a public-key cryptosystem employing QC-MDPC codes, two main applications of ISD exist:

- decoding attacks, that aim at recovering the plaintext from an intercepted ciphertext, which can either be in the form of a syndrome or an error corrupted codeword. In both cases, the corresponding error has weight t, thus an adversary faces a complexity equal to $\frac{C_{\mathsf{ISD}}(n,k,t)}{\sqrt{p}}$, where the polynomial speed-up comes from quasi-cyclicity [33];
- key recovery attacks, that aim at finding low weight codewords in either the public code or its dual. The knowledge about these codewords will indeed reveal the structure of the sparse parity-check matrix used as the private key. In particular, it can be shown that searching for low weight codewords in the dual code corresponds to the optimal attack strategy [8, Section 2.3.1]. We can then assess the complexity of this kind of attacks as $\frac{C_{\mathsf{ISD}}(n_0 p, p, n_0 v)}{p}$.

To reach a security level of λ bits, we must guarantee that all successful attacks run in a time not lower than 2^λ. Hence, taking these considerations into account, we get that v and t must satisfy the following relationships

$$\begin{cases} v \approx \dfrac{\lambda + \log_2(p)}{n_0 \log_2\left(\frac{n_0}{n_0-1}\right)}, \\ t \approx \dfrac{\lambda + \frac{1}{2}\log_2(p)}{\log_2(n_0)}, \end{cases} \tag{1}$$

from which, with simple algebra, we get

$$t \approx v n_0 \left(1 - \frac{\log_2(n_0 - 1)}{\log_2(n_0)} \right) - \frac{\log_2(p)}{2\log_2(n_0)}. \qquad (2)$$

QC-MDPC Codes with $n_0 = 2$. To consider a case of practical interest, we focus on $n_0 = 2$; actually, this corresponds to the QC-MDPC codes that are considered in the BIKE cryptosystem [4] and other relevant works [16,17,34,35]. Assuming $p \approx n_0 v^2$ (recall Definition 6), from (2) we have that

$$\cdot\ t \approx 2v - 0.5 - \log_2(v).$$

For security levels of practical interest, we always have $v < t$: since the resulting QC-MDPC$(2, p, v)$ code always contains codewords of even weight $\leq 2t$ (as stated in Proposition 1), applying Theorem 1 we get that the ML decoder has a provably non-zero DFR. Indeed, we can plug $w = 2v$ into the expression of $\epsilon_{\mathsf{ML}}^{(L)}$, and correspondingly obtain a lower bound on the DFR of the ML decoder as

$$\epsilon_{\mathsf{ML}}^{(L)} = \frac{\binom{2v}{v}\binom{2p-2v}{t-v}}{2\binom{2p}{t}}.$$

Notice that, for growing p and fixed v and t, we got $\epsilon_{\mathsf{ML}}^{(L)} = O\!\left(p^{-v} \right)$, which is polynomial in the circulant size p. To highlight such result, we encapsulate it in the following proposition.

Proposition 2. *Consider $\mathscr{C} \in QC\text{-}MDPC(2, p, v)$ used over a McEliece channel with parameter $t = 2v - 0.5 - \log_2(v)$. Then, ML decoding of $\mathscr{C}$ fails with a probability that decays asymptotically as $O(p^{-v})$.*

This result is foundational, since it proves that, when the parameters v and t are fixed, the DFR of the ML decoder decays polynomially with the circulant size (which is linear in the code length). This is the typical *floor* behavior: the DFR (seen as a function of the code length) starts with an exponential decay but, at some point, the slope changes and the DFR decay becomes only polynomial. To have a further insight on the lower bound of the ML decoder, and to especially highlight how it depends on the code parameters p and v, with simple approximations we elaborate the previous expression and get

$$\epsilon_{\mathsf{ML}}^{(L)} \approx 2^{1.5573v - v\log_2\left(\frac{p}{v}\right) - 0.5\log_2(v) - 1.3257}. \qquad (3)$$

To see how such an estimate has been derived, see Appendix C. To have a graphical view of how $\epsilon_{\mathsf{ML}}^{(L)}$ evolves with the circulant size p, and also to have an evidence of the quality of the approximation in (3), we provide some numerical examples in Fig. 1.

QC-MDPC Codes with $n_0 \geq 4$. Interestingly, for $n_0 \geq 4$, (2) implies $t < v$. Recall that, due to sparsity, we expect that the minimum distance of a large majority of QC-MDPC codes is exactly $2v$. For all such codes, the upper bound $\epsilon_{\mathsf{ML}}^{(U)}$ expressed in Theorem 1 is null, and hence our analysis does not highlight the existence of the floor region.

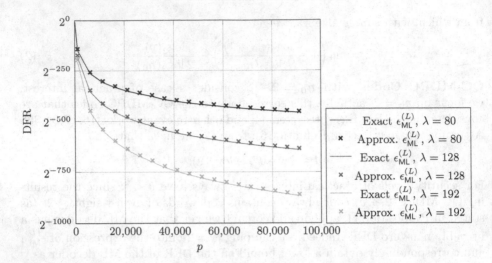

Fig. 1. Lower bound on the DFR of the ML decoder, for QC-MDPC codes with $n_0 = 2$ and parameters achieving different security levels. For each value of p, we have computed v and t through (1). The exact value of $\epsilon_{\mathsf{ML}}^{(L)}$ is computed as in Theorem 1, considering $w = 2v$, while the approximated one has been obtained through (3).

4   Lower Bounds for BF Decoders

As mentioned before, the ML decoder is interesting from a theoretical perspective, since it can be used to derive a safe lower bound on the DFR of any decoder employed in practice. Yet, practical decoders in cryptosystems usually rely on completely different decoding strategies, which originate from the BF decoder first presented in [20]. In this section we propose a numerically-aided approach to compute a lower bound on the DFR of BF decoders. Based on Propositions 3 and 4, we will be able to find error vectors with a special structure by only looking at the code parity-check matrix, without needing any simulation. Then, starting from these error vectors, a lower bound on the DFR of BF decoding can be computed by exploiting some numerical simulations, as will be described in Proposition 5. For this reason, the lower bound we propose, which partially relies on simulations, is defined as numerically-aided.

A BF decoder performs the decoding procedure starting from an estimate of the value of $\mathbf{e}$, initially set to $\mathbf{0}_n$, and changes this estimate, flipping its bit values (hence the name) on the basis of a set of values computed starting from the syndrome, known as *counters*, which are defined as follows.

Definition 9. *Let* $\mathbf{H} \in \mathbb{F}_2^{r \times n}$ *and* $\mathbf{s} = \mathbf{x}\mathbf{H}^\top$, *for* $\mathbf{x} \in \mathbb{F}_2^n$. *We define the i-th counter* σ_i *as the number of set entries in* $\mathbf{s}$ *that are indexed by* $\mathrm{Supp}\,(\mathbf{H}_{:,i})$ *or, equivalently, as the number of unsatisfied parity-check equations in which the i-th bit participates.*

It is immediately seen that almost all QC-MDPC decoders proposed in the literature (like those in [7,16,17,31,34,35,38]) include a stage in which error estimate bit flipping decisions are taken on the basis of counters. So, to encompass all such algorithms, we will generically speak of BF decoders.

Let $\mathbf{x} = \mathbf{c} + \mathbf{e}$, with $\mathbf{c}$ being a codeword and $\mathbf{e}$ being the error vector introduced by the channel. Any BF decoder follows a common procedure, which can be summarized as follows:

1. on input $\mathbf{x} \in \mathbb{F}_2^n$, compute the syndrome $\mathbf{s} = \mathbf{x}\mathbf{H}^\top$ and initialize the error estimate $\mathbf{e}' = \mathbf{0}_n$;
2. compute the counters σ_i, for $i = \{0, \ldots, n-1\}$;
3. assume positions of $\mathbf{e}$ corresponding to high valued counters to be incorrectly estimated, and flip the corresponding entries in $\mathbf{e}$;
4. update the syndrome as $\mathbf{s} + \mathbf{e}'\mathbf{H}^\top$. If the new syndrome is null, complete the procedure outputting $\mathbf{x} + \mathbf{e}'$. If the new syndrome is not null and the maximum number of iterations has not been reached, restart from step 2, otherwise report the occurrence of a failure.

In particular, step 3 is implemented through a threshold criterion: positions associated to counters with values greater than or equal to some threshold $b \leq v$ are considered to be incorrectly estimated. When the decision on a bit is correct (i.e., when the current value of e_i' is different from e_i) we speak of *correct flip*, otherwise (i.e., when the current value of e_i' is equal to e_i) we speak of *wrong flip*. Notice that, in each iteration, we have that $\mathbf{s}$ corresponds to the syndrome of the vector $\mathbf{e} + \mathbf{e}'$. The value of b may be chosen in different ways (for instance, as a function of the iteration number and the syndrome weight), and is not expected to become lower than $v/2$. The reason for this claim is explained next. Indeed, any BF decoder treats as error affected the bits for which the number of unsatisfied involved parity-check equations exceeds that of the satisfied ones. Choosing $b < v/2$ implies that we contradict this criterion, hence we expect that the decoder ends up in performing a number of wrong flips which is larger than that of correct flips.

The counters values are related to the structure of $\mathbf{H}$, as well as to the support of the error vector; the exact relation is described in the next lemma.

Lemma 1. *Let* $\mathbf{H} \in \mathbb{F}_2^{r \times n}$ *and* $\mathbf{s} = \mathbf{e}\mathbf{H}^\top$ *for a vector* $\mathbf{e} \in \mathbb{F}_2^n$. *Let*

$$\gamma_{i,j} = \begin{cases} |\operatorname{Supp}(\mathbf{H}_{:,i}) \cap \operatorname{Supp}(\mathbf{H}_{:,j})| & \text{if } i \neq j, \\ 0 & \text{if } i = j. \end{cases}$$

Let

$$\zeta_i^{(1)}(\mathbf{H}, \mathbf{e}) = \sum_{j \in \operatorname{Supp}(\mathbf{e}) \setminus \{i\}} \gamma_{i,j} - 2 \sum_{\ell \in \operatorname{Supp}(\mathbf{H}_{:,i})} \left\lfloor \frac{\langle \mathbf{H}_{\ell,:}^{(i)} ; \mathbf{e}^{(i)} \rangle}{2} \right\rfloor,$$

$$\zeta_i^{(0)}(\mathbf{H}, \mathbf{e}) = \sum_{j \in \operatorname{Supp}(\mathbf{e})} \gamma_{i,j} - 2 \sum_{\ell \in \operatorname{Supp}(\mathbf{H}_{:,i})} \left\lfloor \frac{\langle \mathbf{H}_{\ell,:} ; \mathbf{e} \rangle}{2} \right\rfloor,$$

where $\mathbf{H}_{\ell,:}^{(i)}$ *and* $\mathbf{e}^{(i)}$ *are the vectors obtained via puncturation of the i-th position. Then, for the i-th counter* σ_i*, the following relation holds*

$$\sigma_i = \begin{cases} \mathrm{wt}(\mathbf{H}_{:,i}) - \zeta_i^{(1)}(\mathbf{H}, \mathbf{e}) & \text{if } e_i = 1, \\ \zeta_i^{(0)}(\mathbf{H}, \mathbf{e}) & \text{if } e_i = 0. \end{cases}$$

Proof. See Appendix D.

4.1 Hard to Decode Errors for QC-MDPC

In this section we rely on Lemma 1 to construct error patterns that, with high probability, cannot be corrected by a BF decoder. Namely, we consider the subset of $B_{n,t}$ formed by the vectors that have a large number of overlapping ones with a column of the parity-check matrix. We show that for such vectors decoding fails with a probability that is rather high, and use numerical simulations to find a lower bound for the DFR of the BF decoder.

Let $\mathscr{C}$ be a QC-MDPC(n_0, p, v) code, with parity-check matrix $\mathbf{H} \in \mathbb{F}_2^{p \times n_0 p}$ and $\mathbf{e} \in B_{n,t}$. As we have already said, a BF decoder takes decisions (i.e., decides which bits are correct and which are error affected) according to the counters values. We expect that high counters are associated to error positions, and low counters are associated to error free positions: if the counters behave in the opposite way (we speak of *bad counters*), then the decoder may make wrong choices. In particular, the decoder may potentially get stuck in a bad configuration when the number of bad counters is rather large. To better explain what we expect to happen in such a situation, let us start with some preliminary considerations.

- Let $\delta(\mathbf{e}) = \max \{\sigma_i \mid i \in \mathrm{Supp}\,(\mathbf{e})\}$. Clearly, a single iteration of a BF decoder with threshold set as $b > \delta(\mathbf{e})$ will never flip any of the set bits in $\mathbf{e}$.
- We expect the same phenomenon happens, with very high probability, even when considering multiple iterations, all employing thresholds larger than $\delta(\mathbf{e})$. Indeed, a flip among the set bits of $\mathbf{e}$ can happen only if the decoder, at some point, makes wrong flips and these flips trigger, in the subsequent iterations, correct flips among the positions indexed by $\mathrm{Supp}\,(\mathbf{e})$. Yet, this phenomenon should happen with extremely low probability. Indeed, when the decoder makes a wrong flip, it moves into a state characterized by more errors: it is very implausible that this somehow helps the decoding process.
- When $\delta(\mathbf{e})$ is particularly low (say, lower than $\lceil v/2 \rceil$), then it is reasonable that decoding fails, regardless of the employed thresholds. Indeed, to flip the set bits in $\mathbf{e}$, a threshold lower than $\lceil v/2 \rceil$ is required. However, with this choice, it becomes very likely that the number of wrong flips exceeds that of correct flips. Hence, the decoder simply increases the overall number of wrongly estimated bits.
- Analogous reasoning can be applied to the case in which an error vector is such that there is a large number of error free positions with high counters values. Indeed, in such a case, the decoder may wrongly flip some of the corresponding bits, and hence will end up in introducing errors.

As we argue in the remainder of this section, finding error vectors leading to bad counters is rather easy for QC-MDPC codes. We start with the following proposition (which can be trivially proven, taking into account that $\mathbf{H}$ is made of circulant blocks).

Proposition 3. *Let* $\mathbf{H} \in \mathbb{F}_2^{p \times n_0 p}$ *be the parity-check matrix of a QC-MDPC code. Then, for any* $\ell \in [0; n-1]$ *and any pair* $i, j \in \mathrm{Supp}\,(\mathbf{H}_{:,\ell})$, *we have* $\gamma_{i,j} \geq 1$.

Remember that, as stated in Lemma 1, high values of $\gamma_{i,j}$ have a bad influence on the counters. Hence, as a consequence of the above proposition, we expect that an error vector whose support is contained in the support of a column of $\mathbf{H}$ leads to large number of bad counters. To formalize this claim, we consider the following proposition.

Proposition 4. *Let* $\mathscr{C} \in QC\text{-}\mathcal{MDPC}(n_0, p, v)$ *with parity-check matrix* $\mathbf{H}$. *Let* $\mathbf{e} \in \mathbb{F}_2^{n_0 p}$ *with weight* $\tilde{t} < v$, *and such that* $\mathrm{Supp}\,(\mathbf{e}) \subseteq \mathrm{Supp}\,(\mathbf{H}_{:,z})$ *for some* z. *Furthermore, assume that*

$$
\begin{cases}
\sum_{\ell \in \mathrm{Supp}(\mathbf{H}_{:,i})} \left\lfloor \dfrac{\langle \mathbf{II}_{\ell,:}^{(i)} \,;\, \mathbf{e}^{(i)} \rangle}{2} \right\rfloor = 0 & \forall i \in \mathrm{Supp}\,(\mathbf{e})\,, \\[2ex]
\sum_{\ell \in \mathrm{Supp}(\mathbf{H}_{:,i})} \left\lfloor \dfrac{\langle \mathbf{H}_{\ell,:} \,;\, \mathbf{e} \rangle}{2} \right\rfloor = 0 & \forall i \in \mathrm{Supp}\,(\mathbf{H}_{:,z}) \setminus \mathrm{Supp}\,(\mathbf{e})\,.
\end{cases}
$$

Then, the following relations hold

$$
\begin{cases}
\sigma_i \leq v + 1 - \tilde{t} & \text{if } i \in \mathrm{Supp}\,(\mathbf{e})\,, \\
\sigma_i \geq \tilde{t} & \text{if } i \in \mathrm{Supp}\,(\mathbf{H}_{:,z}) \setminus \mathrm{Supp}\,(\mathbf{e})\,.
\end{cases}
$$

Proof. The proof is a straightforward application of Lemma 1 and Proposition 3. We start with the case $i \in \mathrm{Supp}\,(\mathbf{e})$ and consider that, by hypothesis, we have $\zeta_i^{(1)}(\mathbf{H}, \mathbf{e}) = \sum_{j \in \mathrm{Supp}(\mathbf{e}) \setminus \{i\}} \gamma_{i,j}$. Since the support of $\mathbf{e}$ is contained in $\mathrm{Supp}\,(\mathbf{H}_{:,z})$, as a consequence of Proposition 3 we have $\gamma_{i,j} \geq 1$ for any pair of indexes $i, j \in \mathrm{Supp}\,(\mathbf{e})$, and hence $\zeta_i^{(1)}(\mathbf{H}, \mathbf{e}) = \sum_{j \in \mathrm{Supp}(\mathbf{e}) \setminus \{i\}} \gamma_{i,j} \geq \tilde{t} - 1$. Then, from Lemma 1 we get $\sigma_i = v - \zeta_i^{(1)}(\mathbf{H}, \mathbf{e}) \leq v + 1 - \tilde{t}$. Analogously, for the case $i \in \mathrm{Supp}\,(\mathbf{H}_{:,z}) \setminus \mathrm{Supp}\,(\mathbf{e})$, we have $\zeta_i^{(0)}(\mathbf{H}, \mathbf{e}) = \sum_{j \in \mathrm{Supp}(\mathbf{e})} \gamma_{i,j} \geq \tilde{t}$, and hence we get $\sigma_i = \zeta_i^{(0)}(\mathbf{H}, \mathbf{e}) \geq \tilde{t}$. □

As an application of the above proposition, we see that increasing $\tilde{t}$ will worsen the counters' behavior: namely, the counters values will become lower for error positions, and higher for the correct positions which are indexed by the column of $\mathbf{H}$ but not by the error vector. In particular, if we choose $\tilde{t} \geq \lceil \frac{v+3}{2} \rceil$, then we will get $\sigma_i \leq \lfloor v/2 \rfloor$ for all $i \in \mathrm{Supp}\,(\mathbf{e})$, and $\sigma_i \geq \lceil \frac{v+3}{2} \rceil$ for all $i \in \mathrm{Supp}\,(\mathbf{H}_{:,z}) \setminus \mathrm{Supp}\,(\mathbf{e})$. To flip the bits indexed by $\mathrm{Supp}\,(\mathbf{e})$, we are going to need a threshold that is not higher than $\lceil v/2 \rceil$, but this will also trigger wrong flips for all positions $i \in \mathrm{Supp}\,(\mathbf{H}_{:,z}) \setminus \mathrm{Supp}\,(\mathbf{e})$. Hence, in such a case, there does not exist a threshold that is sufficiently low to perform correct flips, but also high enough to guarantee that wrong flips do not happen. We point out that an important hypothesis in Proposition 4 is that the values of $\zeta_i^{(0)}(\mathbf{H}, \mathbf{e})$ and $\zeta_i^{(1)}(\mathbf{H}, \mathbf{e})$ only depend on the $\gamma_{i,j}$ values. In general, this is not true and one has to consider also the number of overlapping ones between the error vector and the rows of $\mathbf{H}$. Yet, as we show in the next section, the behavior of the counters remains somehow bad and these vectors cause failures with high probability.

Finally, we comment about the decoding of error vectors with weight $t > v$, but such that their support intersects with the support of a column in $\mathbf{H}$ in a sufficiently large number $\tilde{t}$ of positions. As a difference with the situation we have previously examined, the decoder must now correct more errors. In other words, we can write $\mathbf{e} = \hat{\mathbf{e}} + \check{\mathbf{e}}$, where $\hat{\mathbf{e}}$ and $\check{\mathbf{e}}$ have disjoint supports and $\hat{\mathbf{e}}$ is such that its support has size $\tilde{t}$ and is contained in the support of a column of $\mathbf{H}$. It is very unlikely that these additional errors (i.e., those due to $\check{\mathbf{e}}$) can improve the situation, up to the point that the decoder flips any of the bits in $\hat{\mathbf{e}}$. In the best case scenario, we expect that the decoder may be able to identify the error positions due to $\check{\mathbf{e}}$, but will not be able to flip any of the positions due to $\hat{\mathbf{e}}$. Hence, decoding will fail with very high probability also in this case. Notice that, with a simple counting argument, one finds that the number of errors with weight t and such that their supports intersect in $\tilde{t}$ elements with that of a column in $\mathbf{H}$ (say, the first one) is given by

$$\left|\left\{\mathbf{e} \in B_{n,t}, \text{ such that } |\text{Supp}(\mathbf{e}) \cap \text{Supp}(\mathbf{H}_{:,0})| = \tilde{t}\right\}\right| = \binom{v}{\tilde{t}}\binom{n_0 p - v}{t - \tilde{t}}. \quad (4)$$

In general terms, the possibility to decode successfully depends on many factors (such as the decoder setting) which we have not considered yet. In other words, even if for a vector $\mathbf{e}$ we have $\delta(\mathbf{e}) > \lceil v/2 \rceil$, this does not imply that $\mathbf{e}$ can be corrected. Actually, we expect that a vector with a sufficiently large number of overlapping positions with a column of $\mathbf{H}$ is *"harder to decode"*, with respect to a completely random vector. Hence, even moderately low values of $\tilde{t}$ may lead to rather high decoding failure probabilities. This in turn provides us with an operative method to generate error vector families which are expected to be harder to decode. As a consequence, through the use of numerical simulations to estimate the concrete DFR of these error families, we are able to obtain a lower bound on the DFR of any iterative BF-like decoder, as we state in the following proposition.

Proposition 5 (DFR lower bound).
Let $\mathcal{C} \in \mathcal{QC}\text{-}\mathcal{MDPC}(n_0, p, v)$, with parity-check matrix $\mathbf{H} \in \mathbb{F}_2^{p \times n_0 p}$. Let Dec be a BF-like decoder employed in the McEliece channel with parameter $t > v$, and consider the following procedure:

1. *for any $\tilde{t} \in [1; v]$, generate a large number of vectors with weight t and exactly $\tilde{t} \in [1; v]$ entries that overlap with $\mathbf{H}_{:,0}$;*
2. *simulate decoding of these vectors, and denote with $\tilde{\epsilon}_{\text{Dec}}(\tilde{t})$ the estimated failure rate (that is, the ratio between the number of failure events and that of considered vectors);*

3. compute

$$\epsilon_{\text{Dec}}^{(L)} = \sum_{\tilde{t}=1}^{v} \tilde{\epsilon}_{\text{Dec}}(\tilde{t}) \frac{\binom{v}{\tilde{t}} \binom{n_0 p - v}{t - \tilde{t}}}{\binom{n_0 p}{t}}.$$

Then, $\epsilon_{\text{Dec}}^{(L)}$ represents a lower bound for the DFR of Dec.

Proof. We consider error vectors with a special structure, that is, those intersecting with $\mathbf{H}_{:,0}$ in $\tilde{t} \in [1; v]$ positions. For each $\tilde{t}$, we rely on numerical simulations to estimate the probability that the decoder is not able to correct a vector of this kind, and call this probability $\tilde{\epsilon}_{\text{Dec}}(\tilde{t})$. Assuming that $\tilde{\epsilon}_{\text{Dec}}(\tilde{t})$ is a proper estimate of the failure probability, when considering only vectors $\mathbf{e} \in B_{n,t}$ such that $|\text{Supp}(\mathbf{e}) \cap \text{Supp}(\mathbf{H}_{:,0})| = \tilde{t}$, we have

$$\begin{aligned}
\epsilon_{\text{Dec}}^{(L)} &= \sum_{\tilde{t}=1}^{v} \Pr\left[\text{Dec}(\mathbf{e}) \neq \mathbf{0}_n\right] \cdot \Pr\left[|\text{Supp}(\mathbf{e}) \cap \text{Supp}(\mathbf{H}_{:,0})| = \tilde{t} \mid \mathbf{e} \xleftarrow{\$} B_{n_0 p, t}\right] \\
&\approx \sum_{\tilde{t}=1}^{v} \tilde{\epsilon}_{\text{Dec}}(\tilde{t}) \cdot \Pr\left[|\text{Supp}(\mathbf{e}) \cap \text{Supp}(\mathbf{H}_{:,0})| = \tilde{t} \mid \mathbf{e} \xleftarrow{\$} B_{n_0 p, t}\right] \qquad (5) \\
&= \sum_{\tilde{t}=1}^{v} \tilde{\epsilon}_{\text{Dec}}(\tilde{t}) \cdot \frac{\binom{v}{\tilde{t}} \binom{n_0 p - v}{t - \tilde{t}}}{\binom{n_0 p}{t}},
\end{aligned}$$

where the last equality comes from (4). Finally, we claim that $\epsilon_{\text{Dec}}^{(L)}$ is a lower bound on the DFR since there may be other vectors that cause a decoding failure. For instance, we are not considering vectors that do not intersect with $\mathbf{H}_{:,0}$, but intersect in a large number of positions with other columns of $\mathbf{H}$. For these vectors, we expect to have the same failure rates $\tilde{\epsilon}_{\text{Dec}}^{(L)}$. $\square$

Remark 1. The bound given in the above proposition is likely to be loose. For instance, we may consider the probability that a random $\mathbf{e} \in B_{n,t}$ intersects in $\tilde{t}$ positions with at least a generic column in $\mathbf{H}$. Assuming all columns of $\mathbf{H}$ behave as random vectors with weight v and length p, for rather large values of $\tilde{t}$ we get that such a probability corresponds to

$$1 - \left(1 - \frac{\binom{v}{\tilde{t}} \binom{n_0 p - v}{t - \tilde{t}}}{\binom{n_0 p}{t}}\right)^{n_0 p} \approx n_0 p \frac{\binom{v}{\tilde{t}} \binom{n_0 p - v}{t - \tilde{t}}}{\binom{n_0 p}{t}}.$$

Using these probabilities in (5) (instead of the term $\binom{v}{\tilde{t}} \binom{n_0 p - v}{t - \tilde{t}} / \binom{n_0 p}{t}$), we would obtain an increase on the value of $\epsilon_{\text{Dec}}^{(L)}$ by a factor $n_0 p$. However, this approach leads to multiple counting of the same vectors. We expect that the obtained probabilities are not much higher than the actual ones, yet, using them would prevent us from claiming that $\epsilon_{\text{Dec}}^{(L)}$ is a provable lower bound.

Remark 2. As anticipated in the Introduction, a similar analysis has been independently and concurrently performed by Vasseur in his PhD thesis [41, Chapter

16]. Namely, Vasseur has denoted as *near-codewords* the error patterns producing syndromes with unusually low weight. The effect of near-codewords on the counters distribution has been motivated by the results of numerical simulations, which are reported in [41, Table 16.2]. It can be easily seen that the error vectors we have considered in this section can be deemed as near-codewords, since with very high probability a rather large number of cancellations happen in the syndrome computation. However, as a significant difference with [41], in this paper we have provided a quantitative justification to for the counters behaviours, through Lemma 1 and Propositions 3 and 4.

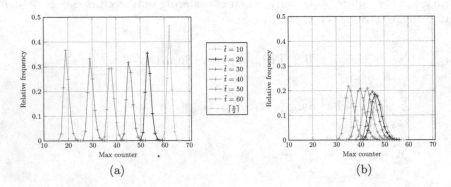

(a)   (b)

Fig. 2. Results of numerical simulations on 100 random codes, picked from the family $QC\text{-}\mathcal{MDPC}(2, p, v)$ with $p = 12,323$ and $v = 71$. For each code, we have generated 100 error vectors intersecting with the first column of $\mathbf{H}$ in $\tilde{t}$ positions. Figures (a) and (b) report the measured distribution of $\max\{\sigma_i \mid i \in \mathrm{Supp}\,(\mathbf{e}) \cap \mathrm{Supp}\,(\mathbf{H}_{:,0})\}$. In (a), we have considered vectors with weight $\tilde{t}$, i.e., such that their support is fully contained in that of $\mathbf{H}_{:,0}$. In (b), we have considered vectors with weight $t = 134$ and support intersecting that of $\mathbf{H}_{:,0}$ in $\tilde{t}$ positions.

4.2   Results for QC-MDPC$(2, p, v)$ Codes

We first consider the counters distribution for error vectors whose support intersects that of a column of $\mathbf{H}$ in $\tilde{t}$ positions. As we have already said, due to overlapping ones with rows of $\mathbf{H}$, we expect the counters values to be slightly better than what we have considered in Proposition 4.

Yet, due to sparsity, we expect that the number of such overlapping elements is low, so that the effect in the counters values is rather limited. To validate this assertion, we have run numerical simulations on the family of QC-MDPC codes with $n_0 = 2$, $p = 12,323$ and $v = 71$, employed in the McEliece channel with $t = 134$. Note that these parameters correspond to the ones of BIKE, version 4.1 [4], considered also in [17]. The obtained results are reported in Fig. 2. We notice that, regardless of the weight of the error vector, when the intersection between the error vector and a column of $\mathbf{H}$ increases, the maximum counter

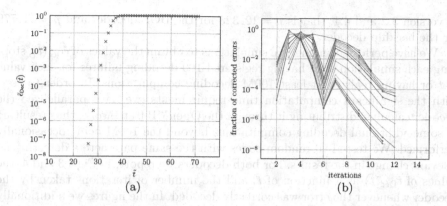

Fig. 3. Numerical simulations on the Backflip decoder as in the BIKE v3.2 specification, with maximum number of iterations set to 100. The sample DFR was estimated running either at least 10^8 decoding actions, or collecting at least 100 decoding failures, whichever event happened first. Figure (b) reports the number of iterations taken to decode an input, for all the inputs which were correctly decoded.

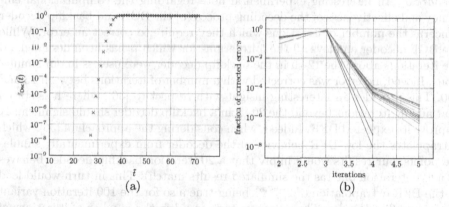

Fig. 4. Numerical simulations on the BGF decoder as in the BIKE v4.1 specification, with maximum number of iterations set to 5. The sample DFR was estimated running either at least 10^8 decoding actions, or collecting at least 100 decoding failures, whichever event happened first. Figure (b) reports the number of iterations taken to decode an input, for all the inputs which were correctly decoded.

becomes lower. Hence, as a consequence, we expect that the failure probability increases, as well.

In order to validate the analysis reported in the previous section, we have applied Proposition 5 on two improved BF decoders, namely, the *backflip* proposed in [35] and the *Black Gray Flip (BGF)* proposed in [17], also used for decryption in BIKE [4]. Both decoders have been considered for codes with $v = 71$ and a McEliece channel with $t = 134$. We have analyzed both decoders for the values of p in the proposals of the BIKE cryptosystem [4], respectively in version 3.2 and 4.1, that is $p = 12,323$ for the BGF decoder and $p = 11,779$ for the backflip decoder.

We have performed numerical simulations to obtain the values of $\tilde{\epsilon}_{\text{Dec}}(\tilde{t})$, stopping each simulation after having registered 100 decoding failures for each value of $\tilde{t}$ or having realized at least 100M decoding computations. In order to cope with the significant computation time requirements, we have parallelized the decoder calculus, distributing it through the OpenMP framework, thus resulting in some additional decoding computations beyond the 100M being occasionally performed. We tested 10 random codes with the same parameters detecting no relevant change in the results. For both decoders we report, in Figs. 3 and 4, the values of $\tilde{\epsilon}_{\text{Dec}}(\tilde{t})$ as a function of $\tilde{t}$, and the number of iterations taken by the decoder whenever the error was correctly decoded. In the figures we additionally report the number of iterations taken by each decoder when a correct decoding computation took place, over the $\approx$ 330M decoded error vectors for each decoder; each colored line reports the data for a specific value of $\tilde{t}$ for which we have determined $\tilde{\epsilon}_{\text{Dec}}(\tilde{t}) > 10^{-8}$.

Remark 3. An interesting experimental note regarding the computational efficiency and effectiveness of the decoding process of the BGF and backflip decoder concerns the number of iterations which they require to correct an error. While the BGF decoder employs all the 5 iterations for which it has been designed, all the iterations above the 12-th in the backflip decoder were useless in our simulations. Indeed, no error was corrected with a number of iterations between 13 and 100. This provides an interesting insight with respect to [36], where it is stated that adding iterations beyond the 20-th in a backflip decoder should significantly improve its expected DFR. Indeed, when considering the approach of [36], which extrapolates the low-DFR behavior of the decoder from experimentally simulable points, our results would imply that a 20 iteration backflip decoder behaves as a 100 iteration one (as the simulated results match). This in turn would lead to the DFR extrapolation of $2^{-97.65}$ being true also for the 100 iteration variant of the backflip decoder. The non monotone trend in the number of iterations of the backflip decoder finds an explanation in the flipping Time-To-Live (TTL) of the procedure (which reverts a bit flip after a given TTL has expired): indeed the TTL during the overwhelming majority of our numerical simulations was found to be set to 5 for flips taking place in the first iteration.

Employing the results on $\tilde{\epsilon}_{\text{Dec}}(\tilde{t})$ obtained through simulations in the expression of $\epsilon_{\text{Dec}}^{(L)}$ given in Proposition 5, we are able to provide lower bounds on the DFR of these algorithms, which are shown in Table 1. The reported values differ from the ones of $\epsilon_{\text{ML}}^{(L)}$ by a factor of $\approx 2^{259}$, in turn showing how a significant amount of failure events, at very low DFR values, are not due to near-codewords (as claimed in [34]). Indeed, our reported data are able to set a reliable lower bound on the DFR, through Proposition 5, in turn showing that iterative decoders perform significantly worse than the ML decoder. We note that the lower bounds we provide do not explicitly contradict the numerical claims on the DFR for both the backflip and the BGF decoder with the param-

Table 1. Summary of the DFR bounds found in this work, compared with the claimed values in [4, 17, 35].

Decoder (p, v, t)	Backflip [35] $(11779, 71, 134)$	BGF [4, 17] $(12323, 71, 134)$
Claimed DFR	2^{-128}	2^{-128}
$\epsilon_{\mathsf{ML}}^{(L)}$	$2^{-425.86}$	$2^{-430.45}$
$\epsilon_{\mathsf{Dec}}^{(L)}$	$2^{-166.3}$	$2^{-168.06}$

eters at hand. We also note that obtaining concrete values for $\tilde{\epsilon}_{\mathsf{Dec}}(\tilde{t})$ for values of $\tilde{t} < 25$ may bring the value of our lower bound further up.

5 Conclusion

We have proposed two approaches for bounding the performance of iterative decoders derived from Gallager's BF, and used in decoding QC-MDPC codes in code-based cryptosystems. The first approach relies on modeling the ML decoder performance, which is an optimal decoder and hence provides an ultimate bound on the behavior of any sub-optimal decoder, such as the BF ones. This also allows to characterize the asymptotic DFR of these decoders, which has been shown to decay polynomially in the code length. The second approach exploits a numerically-aided procedure to provide a lower bound on the DFR of BF decoders: the approach relies on numerical estimations for the DFR of families of error vectors which are harder to decode for BF decoders. Through weighing the contribution to the total DFR of such error families with their size we achieve a lower bound on the DFR for the specific class of iterative decoders derived from BF. In particular, this second approach was shown to provide tighter lower bounds to the DFR by a factor of 2^{259} with respect to the bound obtained modeling the performance of the ML decoder, thus providing a preliminary quantitative assessment of the performance gap of the iterative BF decoders and their ideal ML counterpart on QC-MDPC parameters of interest in code-based cryptography.

Appendix A: Proof of Theorem 1

We focus on ML decoding, and derive an analytical expression for its DFR. To this end, we consider an input $\mathbf{x} = \mathbf{c} + \mathbf{e} \in \mathbb{F}_2^n$, with $\mathbf{c} \xleftarrow{\$} \mathscr{C}$ and $\mathbf{e} \xleftarrow{\$} B_{n,t}$. The decoder first computes $\mathscr{C}^{(\mathbf{x})}$, that is, the set of all codewords that are t away from $\mathbf{x}$, and then outputs at random one of them. Given that, clearly, $\mathbf{c} \in \mathscr{C}^{(\mathbf{x})}$, and that decoding fails every time the decoder output is different from $\mathbf{c}$, we have that a failure happens with probability

$$\frac{\left|\mathscr{C}^{(\mathbf{x})}\right| - 1}{\left|\mathscr{C}^{(\mathbf{x})}\right|}.$$

Note that, if there is only one codeword in $\mathscr{C}^{(\mathbf{x})}$, then this codeword must be $\mathbf{c}$; hence, in this case, we never have a failure. To obtain the DFR, which we denote as ϵ_{ML}, we average the above probability over all the possible errors $\mathbf{e} \in B_{n,t}$, added to all the codewords in $\mathscr{C}$. According to Definition 4, we assume uniform distributions for both $\mathbf{c}$ and $\mathbf{e}$, and hence obtain

$$\epsilon_{\mathsf{ML}} = \frac{1}{2^k \binom{n}{t}} \sum_{\mathbf{e} \in B_{n,t}} \sum_{\mathbf{c} \in \mathscr{C}} \frac{\left|\mathscr{C}^{(\mathbf{c}+\mathbf{e})}\right| - 1}{\left|\mathscr{C}^{(\mathbf{c}+\mathbf{e})}\right|}.$$

Now we show that, due to linearity, we can consider that the transmitted codeword corresponds to $\mathbf{0}_n$. Indeed, for each codeword $\mathbf{c} \in \mathscr{C}$ and any $\mathbf{e} \in B_{n,t}$, we have

$$\mathscr{C}^{(\mathbf{c}+\mathbf{e})} = \{\mathbf{a} \in \mathscr{C} \text{ s.t. } \mathrm{dist}(\mathbf{c}+\mathbf{e}, \mathbf{a}) = t\} = \{\mathbf{a} \in \mathscr{C} \text{ s.t. } \mathrm{dist}(\mathbf{c}+\mathbf{a}, \mathbf{e}) = t\}$$
$$= \{\mathbf{a}' \in \mathscr{C} \text{ s.t. } \mathrm{dist}(\mathbf{a}', \mathbf{e}) = t\} = \mathscr{C}^{(\mathbf{e})}.$$

From this observation, we further obtain

$$\epsilon_{\mathsf{ML}} = \frac{1}{2^k \binom{n}{t}} \sum_{\mathbf{e} \in B_{n,t}} \sum_{\mathbf{c} \in \mathscr{C}} \frac{\left|\mathscr{C}^{(\mathbf{c}+\mathbf{e})}\right| - 1}{\left|\mathscr{C}^{(\mathbf{c}+\mathbf{e})}\right|} = \frac{1}{2^k \binom{n}{t}} \sum_{\mathbf{e} \in B_{n,t}} \sum_{\mathbf{a}' \in \mathscr{C}} \frac{\left|\mathscr{C}^{(\mathbf{e})}\right| - 1}{\left|\mathscr{C}^{(\mathbf{e})}\right|}$$
$$= \frac{1}{2^k \binom{n}{t}} \sum_{\mathbf{e} \in B_{n,t}} 2^k \frac{\left|\mathscr{C}^{(\mathbf{e})}\right| - 1}{\left|\mathscr{C}^{(\mathbf{e})}\right|} = \frac{1}{\binom{n}{t}} \sum_{\mathbf{e} \in B_{n,t}} \frac{\left|\mathscr{C}^{(\mathbf{e})}\right| - 1}{\left|\mathscr{C}^{(\mathbf{e})}\right|}$$
$$= 1 - \frac{1}{\binom{n}{t}} \sum_{\mathbf{e} \in B_{n,t}} \frac{1}{\left|\mathscr{C}^{(\mathbf{e})}\right|}.$$

We now proceed by proving the lower and upper bounds on the DFR. Based on the above aconsiderations, we consider the transmission of the null codeword over the McEliece channel. The output of the channel, which is given as input to the decoder, corresponds to a weight t vector, uniformly distributed over $B_{n,t}$. Decoding fails every time the decoder outputs a codeword which is not the null one. Clearly, for any $\mathbf{e} \in B_{n,t}$, we necessarily have $\mathbf{0}_n \in \mathscr{C}^{(\mathbf{e})}$: hence, a decoding failure may happen only when $\mathscr{C}^{(\mathbf{e})}$ contains at least two codewords. Notice that we can express the decoding failure rate as

$$\epsilon_{\mathsf{ML}} = \frac{1}{\binom{n}{t}} \sum_{\mathbf{e} \in B_{n,t}} \Pr\left[\mathrm{ML}(\mathbf{e}) \neq \mathbf{0}_n\right] = \frac{1}{\binom{n}{t}} \sum_{\mathbf{e} \in B_{n,t}} \sum_{\mathbf{c} \in \mathscr{C} \setminus \mathbf{0}_n} \Pr\left[\mathrm{ML}(\mathbf{e}) = \mathbf{c}\right].$$

Consider that

$$\text{dist}(\mathbf{c}, \mathbf{e}) = \text{wt}(\mathbf{c}) + t - 2\alpha \quad \text{and} \quad \text{dist}(\mathbf{c}, \mathbf{e}) \in [\text{wt}(\mathbf{c}) - t; \text{wt}(\mathbf{c}) + t],$$

where $\alpha = |\text{Supp}(\mathbf{e}) \cap \text{Supp}(\mathbf{c})|$ and, clearly, $0 \leq \alpha \leq \min\{t, \text{wt}(\mathbf{c})\}$. In particular, $\mathbf{c}$ will be at distance t from $\mathbf{e}$ only when $2\alpha = \text{wt}(\mathbf{c})$. Then, the following claims can be straightforwardly proven:

i) if $\text{wt}(\mathbf{c})$ is odd, then $\mathbf{c} \notin \mathscr{C}^{(\mathbf{e})}$;
ii) if $\text{wt}(\mathbf{c}) > 2t$, then $\text{dist}(\mathbf{c}, \mathbf{e}) > t$ and thus $\mathbf{c} \notin \mathscr{C}^{(\mathbf{e})}$;
iii) if $\text{wt}(\mathbf{c})$ is even and $\leq 2t$, then, by a counting argument on the number of elements of $\text{Supp}(\mathbf{e})$ and $\text{Supp}(\mathbf{c})$ that coincide, we have that

$$|\{\mathbf{e} \in B_{n,t} \mid \text{dist}(\mathbf{c}, \mathbf{e}) = t\}| = \binom{\text{wt}(\mathbf{c})}{\text{wt}(\mathbf{c})/2} \binom{n - \text{wt}(\mathbf{c})}{t - \text{wt}(\mathbf{c})/2};$$

iv) if $\mathbf{e}$ is such that $\mathbf{c} \notin \mathscr{C}^{(\mathbf{e})}$, then $\Pr[\text{ML}(\mathbf{e}) = \mathbf{c}] = 0$, otherwise

$$\Pr[\text{ML}(\mathbf{e}) = \mathbf{c}] = \frac{1}{\left| \mathscr{C}^{(\mathbf{e})} \right|} \leq \frac{1}{2},$$

since $\mathscr{C}^{(\mathbf{e})}$ contains at least two codewords.

By putting everything together, we get

$$\epsilon_{\text{ML}} = \frac{1}{\binom{n}{t}} \sum_{\mathbf{c} \in \mathscr{C} \setminus 0_n} \sum_{\mathbf{e} \in B_{n,t}} \Pr[\text{ML}(\mathbf{e}) = \mathbf{c}] = \frac{1}{\binom{n}{t}} \sum_{\mathbf{c} \in \mathscr{C} \setminus 0_n} \sum_{\substack{\mathbf{e} \in B_{n,t} \\ \mathbf{c} \in \mathscr{C}^{(\mathbf{e})}}} \Pr[\text{ML}(\mathbf{e}) = \mathbf{c}]$$

$$= \frac{1}{\binom{n}{t}} \sum_{\mathbf{c} \in \mathscr{C} \setminus 0_n} \sum_{\substack{\mathbf{e} \in B_{n,t} \\ \mathbf{c} \in \mathscr{C}^{(\mathbf{e})}}} \frac{1}{\left| \mathscr{C}^{(\mathbf{e})} \right|} \leq \frac{1}{2\binom{n}{t}} \sum_{\mathbf{c} \in \mathscr{C} \setminus 0_n} |\{\mathbf{e} \in B_{n,t} \mid \text{dist}(\mathbf{c}, \mathbf{e}) = t\}|$$

$$= \frac{1}{2\binom{n}{t}} \sum_{\substack{\mathbf{c} \in \mathscr{C} \setminus 0_n \\ \text{wt}(\mathbf{c}) \in [d; 2t] \\ \text{wt}(\mathbf{c}) \text{ even}}} \binom{\text{wt}(\mathbf{c})}{\text{wt}(\mathbf{c})/2} \binom{n - \text{wt}(\mathbf{c})}{t - \text{wt}(\mathbf{c})/2}$$

$$= \frac{1}{2\binom{n}{t}} \sum_{\substack{w \in [d; 2t] \\ w \text{ even}}} A_w \binom{w}{w/2} \binom{n - w}{t - w/2} = \epsilon_{\text{ML}}^{(U)},$$

where A_w is the number of codewords in $\mathscr{C}$ of weight w, and d is the minimum distance of $\mathscr{C}$.

In analogous way, we now derive a lower bound for the DFR of the ML-decoder; we start from

$$\epsilon_{\mathsf{ML}} = \frac{1}{\binom{n}{t}} \sum_{\mathbf{e} \in B_{n,t}} \Pr\left[\mathsf{ML}(\mathbf{e}) \neq \mathbf{0}_n\right] = \frac{1}{\binom{n}{t}} \sum_{\mathbf{e} \in B_{n,t}} \frac{\left|\mathscr{C}^{(\mathbf{e})}\right| - 1}{\left|\mathscr{C}^{(\mathbf{e})}\right|}$$

$$= \frac{1}{\binom{n}{t}} \sum_{\substack{\mathbf{e} \in B_{n,t} \\ \left|\mathscr{C}^{(\mathbf{e})}\right| \geq 2}} \frac{\left|\mathscr{C}^{(\mathbf{e})}\right| - 1}{\left|\mathscr{C}^{(\mathbf{e})}\right|} \geq \frac{\left|\left\{\mathbf{e} \in B_{n,t} \,\middle|\, \left|\mathscr{C}^{(\mathbf{e})}\right| \geq 2\right\}\right|}{2\binom{n}{t}},$$

where the inequality comes from the observation that, if $\left|\mathscr{C}^{(\mathbf{e})}\right| \geq 2$, we have $\frac{\left|\mathscr{C}^{(\mathbf{e})}\right| - 1}{\left|\mathscr{C}^{(\mathbf{e})}\right|} \geq \frac{1}{2}$. In the above expression, we need to count the number of vectors $\mathbf{e} \in B_{n,t}$ for which $\mathscr{C}^{(\mathbf{e})}$ contains at least a codeword which is different from the null one. To avoid multiple counting of the same vector, we bound further such a quantity as follows. We have

$$\left|\left\{\mathbf{e} \in B_{n,t} \mid \exists \mathbf{c} \in \mathscr{C} \setminus \mathbf{0}_n \text{ s.t. } \operatorname{dist}(\mathbf{c}, \mathbf{e}) = t\right\}\right|$$

$$= \left|\bigcup_{\mathbf{c} \in \mathscr{C} \setminus \mathbf{0}_n} \left\{\mathbf{e} \in B_{n,t} \mid \operatorname{dist}(\mathbf{c}, \mathbf{e}) = t\right\}\right|$$

$$= \left|\left\{\mathbf{e} \in B_{n,t} \mid \operatorname{dist}(\mathbf{c}^*, \mathbf{e}) = t\right\} \cup \left(\bigcup_{\mathbf{c} \in \mathscr{C} \setminus \{\mathbf{0}_n, \mathbf{c}^*\}} \left\{\mathbf{e} \in B_{n,t} \mid \operatorname{dist}(\mathbf{c}, \mathbf{e}) = t\right\}\right)\right|$$

$$\geq \left|\left\{\mathbf{e} \in B_{n,t} \mid \operatorname{dist}(\mathbf{c}^*, \mathbf{e}) = t\right\}\right|,$$

for any non null codeword $\mathbf{c}^*$. Notice that the above quantity depends only on the weight of the considered $\mathbf{c}^*$. Let $w = \operatorname{wt}(\mathbf{c}^*)$: if w is odd or $w \notin [d; 2t]$, then $\left|\left\{\mathbf{e} \in B_{n,t} \mid \operatorname{dist}(\mathbf{c}^*, \mathbf{e}) = t\right\}\right| = 0$, otherwise

$$\left|\left\{\mathbf{e} \in B_{n,t} \mid \operatorname{dist}(\mathbf{c}^*, \mathbf{e}) = t\right\}\right| = \binom{w}{w/2}\binom{n-w}{t-w/2}.$$

Since the above inequality holds for any codeword $\mathbf{c}^*$ of proper weight, we can write $\epsilon_{\mathsf{ML}} \geq \epsilon_{\mathsf{ML}}^{(L)}$, where

$$\epsilon_{\mathsf{ML}}^{(L)} = \max_{\substack{w \in [d; 2t] \\ w \text{ even} \\ A_w > 0}} \left\{\frac{\binom{w}{w/2}\binom{n-w}{t-w/2}}{2\binom{n}{t}}\right\}.$$

Notice that if $\mathscr{C}$ does not contain a codeword with even weight not larger than $2t$, then the expression of $\epsilon_{\mathsf{ML}}^{(L)}$ becomes meaningless (i.e., it becomes 0).

To conclude the proof, we show that the MLS decoder has the same DFR of the ML decoder. Let $\mathbf{x} = \mathbf{c} + \mathbf{e}$, with $\mathbf{c} \in \mathscr{C}$ and $\mathbf{e} \in B_{n,t}$, be the received sequence. The probability that ML, on input $\mathbf{x}$, outputs a codeword which is different from $\mathbf{c}$ is equal to $1 - \left| \mathscr{C}^{(\mathbf{e})} \right|^{-1}$. The MLS decoder, on input $\mathbf{s} = \mathbf{x}\mathbf{H}^{\top}$, fails with probability $1 - \left| S_{\mathbf{H}}^{(\mathbf{x})} \right|^{-1}$. Note that $\mathscr{C}^{(\mathbf{e})}$ contains all codewords $\mathbf{c}' \neq \mathbf{c}$ such that $\mathrm{dist}(\mathbf{e}, \mathbf{c}') = t$; thus, we have that $\mathbf{e}' = \mathbf{c}' + \mathbf{e}$ has weight t and syndrome $\mathbf{e}'\mathbf{H}^{\top} = \mathbf{e}\mathbf{H}^{\top} = \mathbf{s}$, so that $\mathbf{e}' \in S_{\mathbf{H}}^{(\mathbf{x})}$. Then, we have that $\left| \mathscr{C}^{(\mathbf{e})} \right| \leq \left| S_{\mathbf{H}}^{(\mathbf{x})} \right|$.

Now, for each $\mathbf{e}' \in S_{\mathbf{H}}^{(\mathbf{x})}$, we have that $\mathbf{e}\mathbf{H}^{\top} = \mathbf{e}'\mathbf{H}^{\top}$, from which $(\mathbf{e}+\mathbf{e}')\mathbf{H}^{\top} = 0$; hence $\mathbf{c}'' = \mathbf{e} + \mathbf{e}' \in \mathscr{C}$. Now, consider that

$$\mathbf{x} + \mathbf{e}' = \mathbf{c} + \mathbf{e} + \mathbf{e}' = \mathbf{c} + \mathbf{c}'' = \hat{\mathbf{c}} \in \mathscr{C},$$

and that

$$\mathrm{dist}(\hat{\mathbf{c}}, \mathbf{x}) = \mathrm{wt}(\hat{\mathbf{c}} + \mathbf{x}) = \mathrm{wt}(\mathbf{c} + \mathbf{c}'' + \mathbf{c} + \mathbf{e}) = \mathrm{wt}(\mathbf{e} + \mathbf{e}' + \mathbf{e}) = \mathrm{wt}(\mathbf{e}') = t,$$

thus $\hat{\mathbf{c}} \in \mathscr{C}^{(\mathbf{e})}$. This shows that, for any candidate in $S_{\mathbf{H}}^{(\mathbf{x})}$, we also have a candidate in $\mathscr{C}^{(\mathbf{e})}$, and vice versa: this proves that $\left| \mathscr{C}^{(\mathbf{e})} \right| = \left| S_{\mathbf{H}}^{(\mathbf{x})} \right|$.

Appendix B: Proof of Proposition 1

Let $\mathscr{C} \in \mathcal{QC\text{-}MDPC}(n_0, p, v)$, and denote with $\mathbf{H}$ its parity-check matrix formed by circulant blocks of weight v. Let $\mathbf{H}_i$ denote the i-th circulant block in $\mathbf{H}$. For $i_0, i_1 \in \{0, 1, \cdots, n_0 - 1\}$, with $i_0 \neq i_1$, and $\ell \in \{0, 1, \cdots, p - 1\}$, consider a vector $\mathbf{c}^{(i_0, i_1, \ell)}$ in the form

$$\mathbf{c}^{(i_0, i_1, \ell)} = [\mathbf{c}_0^{(i_0, i_1, \ell)}, \mathbf{c}_1^{(i_0, i_1, \ell)}, \cdots, \mathbf{c}_{n_0-1}^{(i_0, i_1, \ell)}],$$

where

$$\mathbf{c}_j^{(i_0, i_1, \ell)} = \begin{cases} \mathbf{0}_p & \text{if } j \neq i_0, i_1, \\ \text{the transpose of the } \ell\text{-th column of } \mathbf{H}_{i_1} & \text{if } j = i_0, \\ \text{the transpose of the } \ell\text{-th column of } \mathbf{H}_{i_0} & \text{if } j = i_1. \end{cases}$$

It is easily seen that $\mathbf{c}^{(i_0, i_1, \ell)}\mathbf{H}^{\top} = \mathbf{0}_p$, hence $\mathbf{c}^{(i_0, i_1, \ell)} \in \mathscr{C}$. Furthermore, $\mathbf{c}^{(i_0, i_1, \ell)}$ has weight $2v$: this proves that $\mathscr{C}$ cannot have a minimum distance larger than $2v$. Consider now that the number of vectors $\mathbf{c}^{(i_0, i_1, \ell)}$ is given by the number of choices for i_0, i_1 and ℓ, which is equal to $p\binom{n_0}{2}$. This proves that $\mathscr{C}$ contains at least $p\binom{n_0}{2}$ codewords with weight $2v$. Clearly, we cannot exclude that there are more codewords with this weight (even if this is rather unlikely), so we can only claim that $A_{2v} \geq p\binom{n_0}{2}$.

Appendix C: Derivation of Equation (3)

We here show how (3) can be obtained. We start by specializing the expression of $\epsilon_{ML}^{(L)}$ for the case of $n_0 = 2$. Remember that the code always contains codewords with weight $w = 2v$, so that we can write

$$\epsilon_{ML}^{(L)} = \frac{\binom{2v}{v}\binom{2p-2v}{t-v}}{2\binom{2p}{t}}$$

For the binomials appearing in the above expression, we are going to use the following well known (for instance, see [15]) approximations

$$\binom{2v}{v} = \frac{2^{2v}}{\sqrt{\pi v}}\left(1 + o(1)\right), \tag{6}$$

$$\binom{2p-2v}{t-v} = \frac{1}{\sqrt{2\pi(t-v)}}\left(\frac{(2p-2v)e}{(t-v)}\right)^{t-v}\left(1 + o(1)\right), \tag{7}$$

$$\binom{2p}{t} = \frac{1}{\sqrt{2\pi t}}\left(\frac{2pe}{t}\right)^{t}\left(1 + o(1)\right), \tag{8}$$

where e is Euler's number. Neglecting the $o(1)$ terms and expressing (6) as a power of 2, we get

$$\binom{2v}{v} \approx 2^{2v - 0.5\log_2(v) - 0.8257}.$$

From (7) and (8), we obtain

$$\frac{\binom{2p-2v}{t-v}}{\binom{2p}{t}} = \frac{1}{\sqrt{1-\frac{v}{t}}}\left(\frac{(2p-2v)e}{(t-v)}\right)^{t-v}\left(\frac{2pe}{t}\right)^{-t}\left(1 + o(1)\right)$$

$$= \frac{e^{-v}}{\sqrt{1-\frac{v}{t}}}\left(\frac{(2p-2v)}{(t-v)}\right)^{t-v}\left(\frac{2p}{t}\right)^{-t}\left(1 + o(1)\right)$$

To further simplify, we consider $t \approx 2v$, from which

$$\frac{\binom{2p-2v}{t-v}}{\binom{2p}{t}} \approx \frac{e^{-v}}{\sqrt{0.5}}\left(\frac{(2p-2v)}{v}\right)^{v}\left(\frac{p}{v}\right)^{-2v}$$

$$= \frac{e^{-v}}{\sqrt{0.5}}\left(\frac{2p}{v} - 2\right)^{v}\left(\frac{p}{v}\right)^{-2v}$$

$$= 2^{-1.4427v + 0.5 + v\log_2\left(\frac{2p}{v} - 2\right) - 2v\log_2\left(\frac{p}{v}\right)}.$$

Since $\frac{2p}{v} \gg 2$, we further have

$$\frac{\binom{2p-2v}{t-v}}{\binom{2p}{t}} \approx 2^{-1.4427v + 0.5 + v\log_2\left(\frac{2p}{v}\right) - 2v\log_2\left(\frac{p}{v}\right)}$$

$$= 2^{-0.4427v + 0.5 - v\log_2\left(\frac{p}{v}\right)}$$

Putting everything together, we get

$$\frac{\binom{2v}{v}\binom{2p-2v}{t-v}}{2\binom{2p}{t}} \approx 2^{1.5573v - v\log_2\left(\frac{p}{v}\right) - 0.5\log_2(v) - 1.3257}$$

Appendix D: Proof of Lemma 1

To avoid confusion, in this proof we use "$\oplus$" to indicate the sum in the binary finite field, and the operator "$+$" to indicate the standard sum over the integers ring. We denote with c_j the value of the j-th parity-check equation $c_j = \bigoplus_{i=0}^{n-1} e_i h_{i,j}$, and we have

$$c_j = 1 \iff \langle \mathbf{H}_{:,j} \, ; \, \mathbf{e} \rangle \text{ is odd.}$$

Recalling that the i-th counter σ_i corresponds to the number of unsatisfied parity-check equations in which the i-th bit participates, that is

$$\sigma_i = \sum_{j \in \text{Supp}(\mathbf{H}_{:,i})} c_j = |\{j \in \text{Supp}(\mathbf{H}_{:,i}) \mid \langle \mathbf{H}_{j,:} \, ; \, \mathbf{e} \rangle \text{ is odd}\}|,$$

whenever $e_i = 1$ we have

$$\sigma_i = \text{wt}(\mathbf{H}_{:,i}) - \left|\left\{j \in \text{Supp}(\mathbf{H}_{:,i}) \mid \langle \mathbf{H}_{j,:}^{(i)}, \mathbf{c}^{(i)} \rangle \text{ is odd}\right\}\right|.$$

We notice that, for each non negative integer a, it results

$$a - 2\left\lfloor \frac{a}{2} \right\rfloor = \begin{cases} 1 & \text{if } a \text{ is odd,} \\ 0 & \text{if } a \text{ is even.} \end{cases}$$

Let $\hbar_{j,\ell}$ denote the lifted entry $h_{i,j}$ (i.e., with value in $\{0;1\} \subseteq \mathbb{Z}$), and consider the following chain of equalities

$$\sigma_i = \left|\left\{j \in \text{Supp}(\mathbf{H}_{:,i}) \mid \langle \mathbf{H}_{j,:}^{(i)} \, ; \, \mathbf{e}^{(i)} \rangle \text{ is odd}\right\}\right|$$

$$= \sum_{j \in \text{Supp}(\mathbf{H}_{:,i})} \langle \mathbf{H}_{j,:}^{(i)}; \mathbf{e}^{(i)} \rangle - 2\left\lfloor \frac{\langle \mathbf{H}_{j,:}^{(i)}; \mathbf{e}^{(i)} \rangle}{2} \right\rfloor$$

$$= \sum_{j \in \text{Supp}(\mathbf{H}_{:,i})} \sum_{\ell \in \text{Supp}(\mathbf{e}^{(i)})} \hbar_{j,\ell} - 2\left\lfloor \frac{\langle \mathbf{H}_{j,:}^{(i)}; \mathbf{e}^{(i)} \rangle}{2} \right\rfloor$$

$$= \left(\sum_{\ell \in \text{Supp}(\mathbf{e})\backslash\{i\}} \sum_{j \in \text{Supp}(\mathbf{H}_{:,i})} \hbar_{j,\ell}\right) - \sum_{j \in \text{Supp}(\mathbf{H}_{:,i})} 2\left\lfloor \frac{\langle \mathbf{H}_{j,:}^{(i)} \, ; \, \mathbf{e}^{(i)} \rangle}{2} \right\rfloor$$

$$= \sum_{\ell \in \text{Supp}(\mathbf{e})\backslash\{i\}} \gamma_{i,\ell} - \sum_{j \in \text{Supp}(\mathbf{H}_{:,i})} 2\left\lfloor \frac{\langle \mathbf{H}_{j,:}^{(i)} \, ; \, \mathbf{e}^{(i)} \rangle}{2} \right\rfloor.$$

Putting all the previous inferences together, the thesis of the Lemma can be easily derived. When $e_i = 0$, the thesis of the Lemma can be proved with analogous reasoning.

References

1. Alagic, G., et al.: Status report on the second round of the NIST post-quantum cryptography standardization process. https://csrc.nist.gov/publications/detail/nistir/8309/final
2. Albrecht, M.R., et al.: Classic McEliece: conservative code-based cryptography. https://classic.mceliece.org/
3. Apon, D., Perlner, R., Robinson, A., Santini, P.: Cryptanalysis of LEDAcrypt. In: Micciancio, D., Ristenpart, T. (eds.) CRYPTO 2020. LNCS, vol. 12172, pp. 389–418. Springer, Cham (2020). https://doi.org/10.1007/978-3-030-56877-1_14
4. Aragon, N., et al.: BIKE: bit flipping key encapsulation. https://bikesuite.org
5. Baldi, M., et al.: A finite regime analysis of information set decoding algorithms. Algorithms **12**(10), 209 (2019)
6. Baldi, M.: QC-LDPC Code-Based Cryptography. Springer Briefs in Electrical and Computer Engineering, Springer, Heidelberg (2014). https://doi.org/10.1007/978-3-319-02556-8
7. Baldi, M., et al.: A failure rate model of bit-flipping decoders for QC-LDPC and QC-MDPC code-based cryptosystems. In: Proceedings of 17th International Joint Conference on e-Business and Telecommunications (ICETE), Secrypt 2020, 17th International Conference on Security and Cryptography, Paris, France, 8–10 July 2020, pp. 238–249 (2020)
8. Baldi, M., et al.: LEDAcrypt: Low-dEnsity parity-check coDe-bAsed cryptographic systems. https://www.ledacrypt.org/
9. Baldi, M., et al.: Security of generalised Reed-Solomon code-based cryptosystems. IET Inf. Secur. **13**(4), 404–410 (2019)
10. Becker, A., Joux, A., May, A., Meurer, A.: Decoding Random Binary Linear Codes in $2^{n/20}$: How $1+1 = 0$ improves information set decoding. In: Pointcheval, D., Johansson, T. (eds.) EUROCRYPT 2012. LNCS, vol. 7237, pp. 520–536. Springer, Heidelberg (2012). https://doi.org/10.1007/978-3-642-29011-4_31
11. Berlekamp, E.R., McEliece, R.J., van Tilborg, H.C.A.: On the inherent intractability of certain coding problems. IEEE Trans. Inf. Theory **24**(3), 384–386 (1978)
12. Both, L., May, A.: Decoding linear codes with high error rate and its impact for LPN security. In: Lange, T., Steinwandt, R. (eds.) PQCrypto 2018. LNCS, vol. 10786, pp. 25–46. Springer, Cham (2018). https://doi.org/10.1007/978-3-319-79063-3_2
13. Canto-Torres, R., Tillich, J.: Speeding up decoding a code with a non-trivial automorphism group up to an exponential factor. In: Proceedings IEEE International Symposium on Information Theory (ISIT 2019), Paris, France, 7–12 July 2019, pp. 1927–1931 (2019)
14. Couvreur, A., et al.: Distinguisher-based attacks on public-key cryptosystems using Reed-Solomon codes. Designs Codes Crypt. **73**(2), 641–666 (2014). https://doi.org/10.1007/s10623-014-9967-z
15. Das, S.: A brief note on estimates of binomial coefficients (2016). http://page.mi.fu-berlin.de/shagnik/notes/binomials.pdf
16. Drucker, N., Gueron, S.: A toolbox for software optimization of QC-MDPC code-based cryptosystems. J. Cryptogr. Eng. **9**(4), 341–357 (2019). https://doi.org/10.1007/s13389-018-00200-4
17. Drucker, N., Gueron, S., Kostic, D.: QC-MDPC decoders with several shades of gray. In: Ding, J., Tillich, J.-P. (eds.) PQCrypto 2020. LNCS, vol. 12100, pp. 35–50. Springer, Cham (2020). https://doi.org/10.1007/978-3-030-44223-1_3

18. Eaton, E., Lequesne, M., Parent, A., Sendrier, N.: QC-MDPC: a timing attack and a CCA2 KEM. In: Lange, T., Steinwandt, R. (eds.) PQCrypto 2018. LNCS, vol. 10786, pp. 47–76. Springer, Cham (2018). https://doi.org/10.1007/978-3-319-79063-3_3
19. Faugère, J.-C., Otmani, A., Perret, L., Tillich, J.-P.: Algebraic cryptanalysis of McEliece variants with compact keys. In: Gilbert, H. (ed.) EUROCRYPT 2010. LNCS, vol. 6110, pp. 279–298. Springer, Heidelberg (2010). https://doi.org/10.1007/978-3-642-13190-5_14
20. Gallager, R.G.: Low-Density Parity-Check Codes. M.I.T Press, Cambridge (1963)
21. Guo, Q., Johansson, T., Stankovski, P.: A key recovery attack on MDPC with CCA security using decoding errors. In: Cheon, J.H., Takagi, T. (eds.) ASIACRYPT 2016. LNCS, vol. 10031, pp. 789–815. Springer, Heidelberg (2016). https://doi.org/10.1007/978-3-662-53887-6_29
22. Hofheinz, D., Hövelmanns, K., Kiltz, E.: A modular analysis of the Fujisaki-Okamoto transformation. In: Kalai, Y., Reyzin, L. (eds.) TCC 2017. LNCS, vol. 10677, pp. 341–371. Springer, Cham (2017). https://doi.org/10.1007/978-3-319-70500-2_12
23. Khathuria, K., Rosenthal, J., Weger, V.: Encryption scheme based on expanded Reed-Solomon codes. Adv. Math. Commun. **15**(2), 207–218 (2021)
24. Löndahl, C., Johansson, T., Koochak Shooshtari, M., Ahmadian-Attari, M., Aref, M.R.: Squaring attacks on McEliece public-key cryptosystems using quasi-cyclic codes of even dimension. Designs Codes Crypt. **80**(2), 359–377 (2015). https://doi.org/10.1007/s10623-015-0099-x
25. McEliece, R.J.: A public-key cryptosystem based on algebraic coding theory. In: DSN Progress Report, pp. 114–116 (1978)
26. National Institute of Standards and Technology. NIST Post-Quantum Standardization Process (2017). https://csrc.nist.gov/Projects/Post-Quantum-Cryptography
27. Ouzan, S., Be'ery, Y.: Moderate-density parity-check codes (2009). https://arxiv.org/abs/0911.3262
28. Poltyrev, G.: Bounds on the decoding error probability of binary linear codes via their spectra. IEEE Trans. Inf. Theory **40**(4), 1284–1292 (1994)
29. Misoczki, R., Tillich, J.-P., Sendrier, N., Barreto, P.S.L.M.: MDPC-McEliece: new McEliece variants from moderate density parity-check codes. In: Proceedings of IEEE International Symposium on Information Theory (ISIT 2013). Istanbul, Turkey, 7–12 July 2013, pp. 2069–2073 (2013)
30. Santini, P., Battaglioni, M., Chiaraluce, F., Baldi, M.: Analysis of reaction and timing attacks against cryptosystems based on sparse parity-check codes. In: Baldi, M., Persichetti, E., Santini, P. (eds.) CBC 2019. LNCS, vol. 11666, pp. 115–136. Springer, Cham (2019). https://doi.org/10.1007/978-3-030-25922-8_7
31. Santini, P., et al.: Analysis of the error correction capability of LDPC and MDPC codes under parallel bit-flipping decoding and application to cryptography. IEEE Trans. Commun. **68**(8), 4648–4660 (2020)
32. Santini, P., et al.: Hard-decision iterative decoding of LDPC codes with bounded error rate. In: Proceedings of IEEE International Conference on Communications (ICC 2019), Shanghai, China, 20–24 May 2019 (2019)
33. Sendrier, N.: Decoding one out of many. In: Yang, B.-Y. (ed.) PQCrypto 2011. LNCS, vol. 7071, pp. 51–67. Springer, Heidelberg (2011). https://doi.org/10.1007/978-3-642-25405-5_4
34. Sendrier, N., Vasseur, V.: About low DFR for QC-MDPC decoding. In: Ding, J., Tillich, J.-P. (eds.) PQCrypto 2020. LNCS, vol. 12100, pp. 20–34. Springer, Cham (2020). https://doi.org/10.1007/978-3-030-44223-1_2

35. Sendrier, N., Vasseur, V.: On the decoding failure rate of QC-MDPC bit-flipping decoders. In: Ding, J., Steinwandt, R. (eds.) PQCrypto 2019. LNCS, vol. 11505, pp. 404–416. Springer, Cham (2019). https://doi.org/10.1007/978-3-030-25510-7_22
36. Sendrier, N., Vasseur, V.: On the existence of weak keys for QC-MDPC decoding. Cryptology ePrint Archive, Report 2020/1232 (2020). https://eprint.iacr.org/2020/1232
37. Sidelnikov, V.M., Shestakov, S.O.: On insecurity of cryptosystems based on generalized Reed-Solomon codes. Discret. Math. Appl. **2**(4), 439–444 (1992)
38. Tillich, J.-P.: The decoding failure probability of MDPC codes. In: Proceedings of IEEE International Symposium on Information Theory (ISIT 2018), Vail, CO, USA, 17–22 June 2018, pp. 941–945 (2018)
39. Canto Torres, R., Sendrier, N.: Analysis of information set decoding for a sub-linear error weight. In: Takagi, T. (ed.) PQCrypto 2016. LNCS, vol. 9606, pp. 144–161. Springer, Cham (2016). https://doi.org/10.1007/978-3-319-29360-8_10
40. Vardy, A.: The intractability of computing the minimum distance of a code. IEEE Trans. Inf. Theory **43**(6), 1757–1766 (1997)
41. Vasseur, V.: Post-quantum cryptography: study on the decoding of QC-MDPC codes. Ph.D. thesis (2021)

Author Index

Printed in the United States
by Baker & Taylor Publisher Services

Printed in the United States
by Baker & Taylor Publisher Services